International Scientific and Technical Conference "Innovative Mechanical Engineering Technologies, Equipment and Materials-2014"

IOP Conference Series: Materials Science and Engineering
Volume 86

Kazan, Russian Federation
3-5 December 2014

ISBN: 978-1-5108-0750-1
ISSN: 1757-8981

Printed from e-media with permission by:

Curran Associates, Inc.
57 Morehouse Lane
Red Hook, NY 12571

Some format issues inherent in the e-media version may also appear in this print version.

Printed by Curran Associates, Inc. (2015)

For permission requests, please contact the Institute of Physics
at the address below.

Institute of Physics
Dirac House, Temple Back
Bristol BS1 6BE UK

Phone: 44 1 17 929 7481
Fax: 44 1 17 920 0979

techtracking@iop.org

Additional copies of this publication are available from:

Curran Associates, Inc.
57 Morehouse Lane
Red Hook, NY 12571 USA
Phone: 845-758-0400
Fax: 845-758-2634
Email: curran@proceedings.com
Web: www.proceedings.com

Table of contents

Volume 86

International Scientific and Technical Conference "Innovative Mechanical Engineering Technologies, Equipment and Materials-2014"
3–5 December 2014, Kazan Federal University, Kazan, Russian Federation

Accepted papers received: 20 May 2015
Published online: 26 June 2015

Preface

Papers

IOP Conf. Series: Materials Science and Engineering **86** (2015) 011001 doi:10.1088/1757-899X/86/1/011001

Preface

In the period from 3 to 5 December 2014 the city of Kazan hosted the International Scientific Conference "Innovative mechanical engineering technologies, equipment and materials – 2014" (ISC "IMETEM – 2014"). The event was followed by the 14th International specialized exhibition "Engineering. Metalworking. Kazan"

The main objective of the annual conference was for participants to discuss scientific and technical achievements in the design and manufacture of engineering products, the expansion of cooperation between scientific organizations and enterprises of machine-building complex and the definition of perspective ways of creation and development of new techniques, technologies and materials.

The conference "IMETEM" was devoted to the 90th anniversary of Fayzrahman Salahovich Yunusov, who made a great contribution in the field of aviation technology.

Kashapov Nail, D.Sc., professor (Kazan Federal University)

IOP Conf. Series: Materials Science and Engineering **86** (2015) 011002 doi:10.1088/1757-899X/86/1/011002

Peer review statement

All papers published in this volume of *IOP Conference Series: Materials Science and Engineering* have been peer reviewed through processes administered by the proceedings Editors. Reviews were conducted by expert referees to the professional and scientific standards expected of a proceedings journal published by IOP Publishing.

New approaches for evaluating rheological models in composites

A N Bobryshev[1], A V Lakhno[1], P V Voronov[2], E R Galimov[3], R F Sharafutdinov[3] and N Ya Galimova[3]

[1] Penza State University of Architecture and Construction, Russia
[2] Non-state educational private institution CPE "Institute "Profi"", Russia
[3] Kazan National Research Technical University named after A N Tupolev, Russia

E-mail: kstu-material@mail.ru

Abstract. The classic rheological models constructed of elastic and viscous elements are briefly analyzed. A model of relaxation mechanism in composites is proposed.

1. Introduction

Kinetic processes are of great importance in materials science since they determine the stages and duration of structure formation influencing the final state of the materials and products based on them. In operating practice they appear in the form of creep, relaxation, shrinkage, swelling and etc. However both technological and operating parameters are determined by the state of the material structure under study. In this paper the classic rheological models constructed of elastic and viscous elements are analyzed to evaluate kinetic processes in composites [1-4].

It is well known that when developing the new composite materials the stress relaxation is of great importance because it is directly related to the response of the material structure to the prolonged mechanical impact in the mode of constant strain [3, 4]. Kinetic regulations of the relaxation in classical consideration have trivial exponential solution which of the does not correspond to the experimental values. More complicated empirical regularities which are however based on the exponential function are used for adequate interpretation of the experimental data.

In this situation quite an urgent task is to search for the new kinetic models that provide noncontradictory evaluation of these phenomena and reveal their nature [4].

2. Governing equations

We will briefly analyze the classical rheological models which provide more consistent and reasoned transition to the developed model of relaxation mechanism. Composite materials show tendency to viscous flow under long exposure on them of applied load. When the load is removed due to the display of elastic properties part of the strain is instantly restored. Afterwards the remaining rubbery deformation gradually decreases. Therefore under load composite materials exhibit viscoelastic properties which are considered from the perspective of rheology.

In rheology behavior of materials is most often described by the mechanical models constructed of elastic and viscous elements. Elastic structural elements are simulated by a spring in which the relationship between stress and strain is linear according to Hooke's law:

$$\sigma_y = E\varepsilon_y, \tag{1}$$

where E – modulus of elasticity, ε_y – elastic strain.

Viscous element is represented as a fluid damping device its behavior is described by the Newtonian flow law.

$$\sigma_b = \eta \mathrm{d}\varepsilon_b / \mathrm{d}t, \tag{2}$$

where η – viscosity coefficient, ε_b – viscous strain. Various combinations of elastic and viscous elements allow making efficient modeling of rheological processes.

Stress relaxation process in the elementary representation is described by the Maxwell model with the simplest (consecutive) combination of elastic and viscous elements. A distinctive feature of this model is in the fact that with increasing strain viscous properties of Maxwell's model element become exhausted and disappear while elastic properties are fully kept.

Deforming the model on a certain finite value ε leads to the instantaneous elastic strain ε_y and gradual development of viscous strain ε_b in time. Consequently the total strain is determined according to the principle of additivity taking into account the model structure

$$\varepsilon = \varepsilon_y + \varepsilon_b. \tag{3}$$

The rate of change of the strain is given by the expression

$$\mathrm{d}\varepsilon / \mathrm{d}t = (\mathrm{d}\varepsilon_y / \mathrm{d}t) + (\mathrm{d}\varepsilon_b / \mathrm{d}t). \tag{4}$$

Using (1) and (2) we get $(\mathrm{d}\varepsilon_y / \mathrm{d}t) = (1/E)(\mathrm{d}\sigma / \mathrm{d}t)$ and $(\mathrm{d}\varepsilon_b / \mathrm{d}t) = \sigma / \eta$. Taking into account the above equations dependence (4) takes the form

$$\mathrm{d}\varepsilon / \mathrm{d}t = (1/E)(\mathrm{d}\sigma / \mathrm{d}t) + \sigma / \eta. \tag{5}$$

According to the condition at the initial time $t = 0$ strain takes the value $\varepsilon = \varepsilon_0$ and does not change subsequently $(\mathrm{d}\varepsilon/\mathrm{d}t) = 0$. This allows you to integrate equation (5)

$$\sigma = \sigma_0 \exp(-t/\tau_p), \tag{6}$$

where $\sigma_0 = E\varepsilon_0$ – initial stress, $\tau_p = \eta/E$ – relaxation time. The obtained widely known relation characterizes the kinetics of the stress relaxation process.

The behavior of real materials is described by a more complex model studied by Kuhn in which the set of Maxwell elements with different values of constants E and η are connected in parallel. It is obvious that the relaxation properties of the constituent elements are different. Because of the uncertainty of particular values of E and η for individual Maxwell elements the study of such a model is quite difficult. However Kuhn's multi-element model allows you to reveal the relaxation mechanism more fully. In this regard we consider the deformation of a heterogeneous composite material. In composite materials the elastic and viscous structural elements are in the different states of deformation. A model of this heterogeneous system can be represented as the set of parallel-connected Maxwell elements with various degrees of deformity (figure 1).

Suppose that elasticity E and viscosity η coefficients for all the elements are identical. This conditional averaging moves the model away from a real material but allows you to give qualitatively new evaluation of the stress relaxation process.

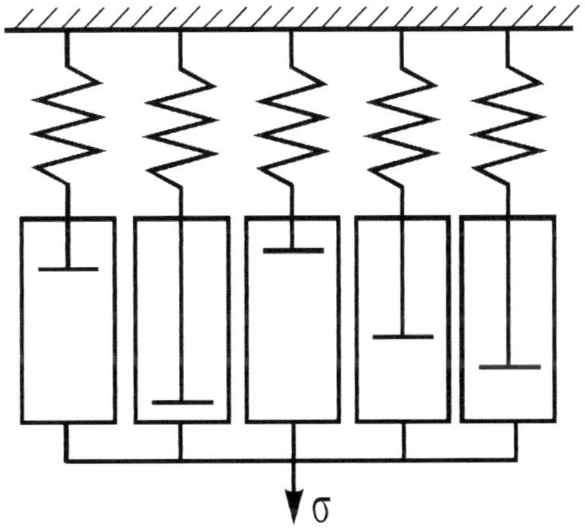

Figure 1. Multi-element model of stress relaxation (Kuhn model).

Most simply the new quality can be seen when considering the following example. Let us analyze the two multi-element models. Each model consists of five Maxwell elements connected in parallel and is stretched by the applied stress to a fixed value of the relative strain (figure 2). Values of E and η for all elements of the models are identical. In the first one viscous properties are exhausted in the four elements and only one has a reserve of viscosity (figure 2*a*). In the second model we can see the opposite situation (figure 2*b*).

In this regard the difference of stress relaxation time period is of interest. From the intuitive considerations it follows that since the second model has the higher reserve of viscous properties then relaxation will occur over a longer time.

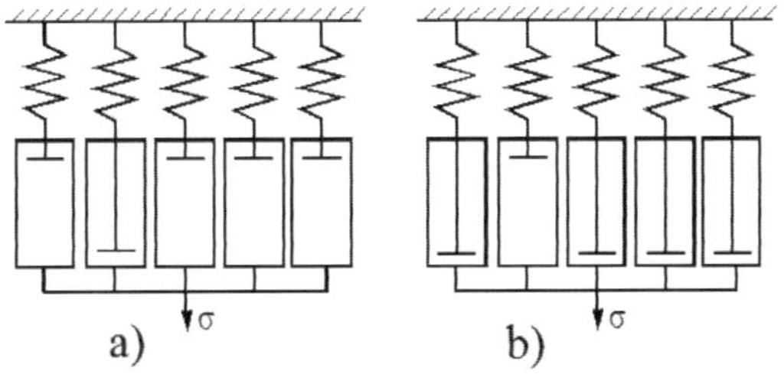

Figure 2. Kuhn's multi-elemental model.

It should be noted that viscous properties of both models can be characterized by a single Newtonian equation [4] since coefficient E and η for all elements are identical. Then however the degree of initial deformity of the elements should be taken into account. In this representation we can adapt the Newtonian equation if in the left-hand side we will note the fraction of Maxwell elements n_b with reserve of viscous properties

$$n_b\sigma = \eta \mathrm{d}\varepsilon_b / \mathrm{d}t . \quad (7)$$

In heterogeneous materials value n_b is equivalent to the volume fraction with the reserve of viscous properties. Elements of this volume are distributed in the total volume of the material under study in an arbitrary way. Amount of volume with the reserve of viscous properties depend on the production technology of material and its operating conditions.

Topology of the distribution of viscous and hard phases can be represented by the Shklovsky-de Gennes model in which a representative volume is filled with a viscous phase permeated with a spatially connected rigid structural framework [5]. Without describing the structure in detail the framework can be presented in the form of spatial distorted lattice with characteristic cells, their averaged size is l_τ. With increasing the volume of viscous phase the rigid framework becomes sparser and the size of its cells l_c increases and vice versa. Thus the following condition $n_b \propto l_\tau$ is satisfied. Taking this into account equation (7) takes the form

$$l_c\sigma = \eta \mathrm{d}\varepsilon_b/\mathrm{d}t. \tag{8}$$

This implies a very important conclusion that Newtonian equation (2) describes the viscous properties only of the homogeneous systems with identical deformity of the unit elements and is a special case of a more general relation (8). A distinctive feature of equation (8) lies in the fact that via conditional size l_c of the volume with hard properties it takes into account the degree of heterogeneity that directly affects the kinetics of stress relaxation in the material.

Let us analyze the kinetics of stress relaxation taking into account equation (8). First of all we note that in the relaxation process the volume with a reserve of viscous properties hyperbolically decreases over time

$$V_b = C_b/t, \tag{9}$$

where C_b – constant. Volume V_b and characteristic structural size l_c are interconnected by the dependency

$$V_b = l_c^{D_i} = C_b/t, \tag{10}$$

where D_i – structural dimension of a heterogeneous system. From the relationship (10) we find

$$l_c = C_b^{1/D_i} t^{(-1/D_i)}. \tag{11}$$

Substituting (11) into equation (8) we get

$$C_b^{1/D_i} t^{(-1/D_i)} \sigma = \eta \mathrm{d}\varepsilon_b/\mathrm{d}t. \tag{12}$$

Taking into account (12) equation (5) takes the form

$$\mathrm{d}\varepsilon/\mathrm{d}t = (1/E)(\mathrm{d}\sigma/\mathrm{d}t) + (\sigma/\eta)C_b^{1/D_i} t^{(-1/D_i)}.$$

Taking the initial conditions $t = 0$, $\varepsilon = \varepsilon_0$, $(\mathrm{d}\varepsilon/\mathrm{d}t) = 0$ into consideration we get

$$(1/E)(\mathrm{d}\sigma/\mathrm{d}t) + (\sigma/\eta)C_b^{1/D_i} t^{(-1/D_i)} = 0.$$

Separating the variables we find

$$\mathrm{d}\ln\sigma = -(E/\eta)C_b^{1/D_i}[1-(1/D_i)]^{-1}\mathrm{d}t^{[1-(1/D_i)]}.$$

Integrating the obtained equality in the limits $(0 \div t)$ and $(\sigma_0 \div \sigma)$ we obtain

$$\sigma = \sigma_0 \exp\left[-\alpha_b t^{1-(1/D_i)}\right], \tag{13}$$

where $-\alpha_b = (E/\eta)C_b^{1/D_i}[1-(1/D_i)]^{-1}$ – constant. Relationship (13) can be rewritten in a simpler form

$$\sigma = \sigma_0 \exp\left[-\alpha_b t^n\right], \tag{14}$$

where $n = 1 - (1/D_i)$ – constant for a given heterogeneous system $\alpha_b = \tau_p^{-n}$.

We note that analytical equation (14) corresponds to an empirical equation of V.M. Ravinsky for metals [6].

For polymers along with elastic (reversible) and plastic (irreversible) strain the reversible rubbery strain can also be observed it is most evident when removing the strain in elastomers. In this case the Maxwell model becomes inapplicable. The Zener model in which Maxwell element is connected in parallel with the elastic element is most common and well known to adequately describe the rheological behavior of such materials.

It should be noted that Zener model can be easily converted either in the Maxwell model (at $E_1 = 0$) or the Kelvin-Voigt model (at $E_2 = \infty$). Let us rewrite the equation (5) for the Maxwell model in the following form

$$\eta(\mathrm{d}\varepsilon/\mathrm{d}t) = (\eta/E)(\mathrm{d}\sigma/\mathrm{d}t) + \sigma.$$

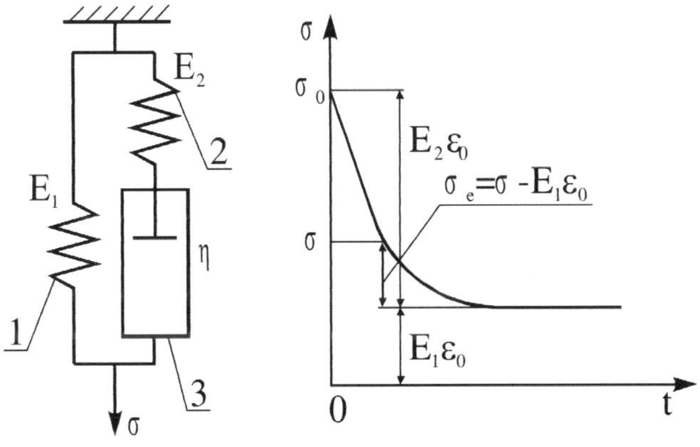

Figure 3. Zener model: 1 – elastic Zener element; 2 – elastic Maxwell element; 3 – viscous element.

Then differential equation corresponding to the Zener model will have the form

$$E_1\varepsilon_0 + \eta(\mathrm{d}\varepsilon/\mathrm{d}t) = (\eta/E_2)(\mathrm{d}\sigma/\mathrm{d}t) + \sigma, \quad (15)$$

where E_1 – equilibrium modulus of elasticity (figure 3). In the stress relaxation process the initial strain is kept constant during the whole time $\varepsilon = \varepsilon_0 = \mathrm{const}$, $\mathrm{d}\varepsilon/\mathrm{d}t = 0$. Taking the initial conditions into consideration we transform (15)

$$E_1\varepsilon_0 - \sigma = (\eta/E_2)(\mathrm{d}\sigma/\mathrm{d}t) + \sigma. \quad (16)$$

We introduce a new variable $z = \sigma - E_1\varepsilon_0$, $\mathrm{d}z = \mathrm{d}\sigma$, taking it into account we obtain the solution of equation (16)

$$z = z_0 \exp[-(E_2/\eta)t], \quad (17)$$

where $z_0 = (\sigma_0 - E_1\varepsilon_0)$ – constant, σ_0 – stress at the initial time (at $t = 0$). Let us represent relationship (17) in the final form

$$\sigma = E_1\varepsilon_0 + (\sigma_0 - E_1\varepsilon_0)\exp[-(E_2/\eta)t]. \quad (18)$$

We express the value of σ_0 according to Hooke's law

$$\sigma_0 = E_0\varepsilon_0, \quad (19)$$

where E_0 – modulus of instantaneous elasticity for most of the polymers it is ~ 10^3 MPa.

Taking into account the latter equality we simplify expression (18)

$$\sigma = E_1\varepsilon_0 + E_2\varepsilon_0 \exp[-(E_2/\eta)t], \quad (20)$$

where $E_2 = (E_0 - E_1)$ – modulus of elasticity of Maxwell element in the Zener model.

Introduced variable $z = \sigma - E_1\varepsilon_0 = \sigma_e(\mathrm{d}z = \mathrm{d}\sigma_e)$ in equation (16) is substantially the part (σ_e) of the total stress σ which undergoes relaxation (figure 3). Therefore taking into account the topology of the distribution of viscous and hard phases equation (16) can be written in another form

$$l_c\sigma_e = -(\eta/E_2)(\mathrm{d}\sigma/\mathrm{d}t). \quad (21)$$

Solving equation (21) using dependency (11) we finally find

$$\sigma = E_1\varepsilon_0 + E_2\varepsilon_0 \exp(-\alpha_b t^n), \quad (22)$$

where $\alpha_b = (E_2/\eta)C_b^{1/D_i}[1-(1/D_i)]^{-1}$, $E_1\varepsilon_0 = \sigma_\infty$ – equilibrium stress. It is noteworthy that the resulting equation (22) completely coincides with the empirical Kohlrausch formula studied by the G.L. Slonimsky with colleagues with regard to polymers [7].

Value of empirical measure n is estimated by the interval $n = 0 \div 1$ which corresponds to the equality $n = 1 - (1/D_i)$ where dimension of D_i taking into account the occurrence of fractality in the structure of heterogeneous material theoretically can be in the range $1 < D_i < \infty$. Thus change in the measure n is determined by the fractal variability of the structure of heterogeneous material this allows you to reveal the physical nature of the stress relaxation process more deeply. Defining experimentally the measure n and determining values of structural dimensions D_i and D_e you can make the topological evaluation of state of distribution of the effective relaxation zones in the material structure.

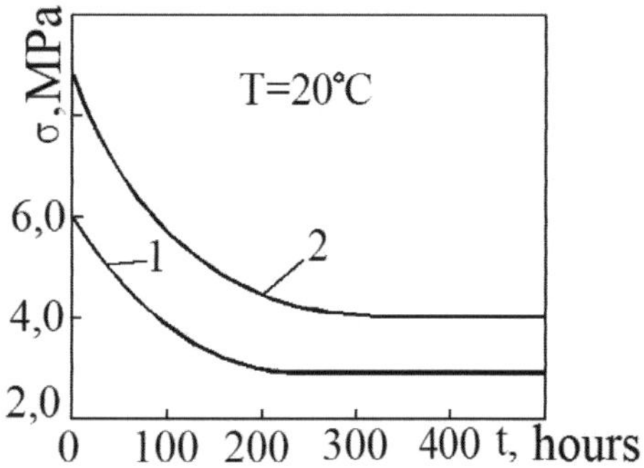

Figure 4. Stress relaxation in an epoxy-filled composite (1 – σ_0 = 6 MPa, σ_∞ = 2.9 MPa; 2 – σ_0 = 9 MPa, σ_∞ = 4 MPa).

Experimental data on stress relaxation in epoxy-filled composite (see figure 4) show that the depth ($\sigma_0 - \sigma_\infty$) of relaxation increases with the growth of initial σ_0 stress (at t = 0). In this case according to dependency (22) coefficient α_b does not change (α_b = 7.4 hour^{-n}) and measure n increases. Here the change in the value of measure n is of interest.

When initial stress σ_0 = 6 MPa, measure n is equal to 0.47 and external dimension of relaxation areas is determined by the value D_e = 1.94. In turn when σ_0 = 9 MPa, measure n is equal to 0.51 and this corresponds to the dimension D_e = 2.02. Consequently, dimension of relaxation areas increases with the growth of initial stress. If dimension D_e = 1.94 determines the fractal between line and surface then dimension D_e = 2.02 is related to the relaxation zones having the surface topology. It is obvious that areas with higher dimension provide deeper relaxation (see figure 4).

Depth of the relaxation depends significantly on the external environment. Figure 5 illustrates the relaxation dependences of polymethylmethacrylate (PMMA) in various environments. From these experiments we can see that environmental activity with respect to the material increases the completeness of stress relaxation like the increase of the initial stress σ_0 (figure 4). For PMMA environmental activity increases in the series water – kerosene – acetone – benzene. Environments – solvents (acetone and benzene) provide the most efficient relaxation related to both crack formation and release of molecular mobility.

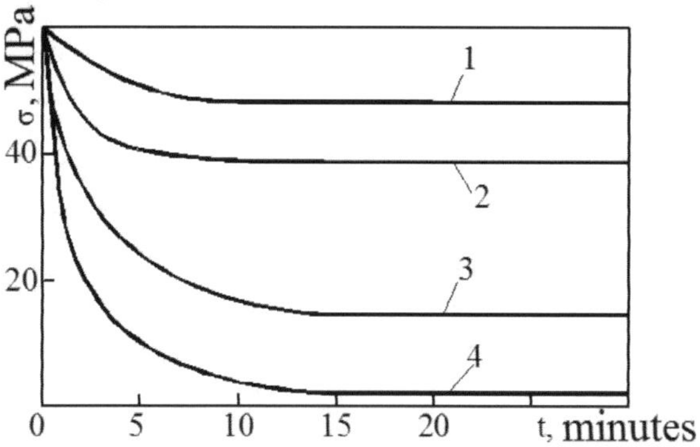

Figure 5. Stress relaxation in PMMA depending on the environment: 1 – water, 2 – kerosene, 3 – acetone, 4 – benzene.

In this case coefficient α_b in equation (22) also remains constant (α_b = 1.8 min^{-n}) and external dimension D_e of relaxation areas increases with the growth of environmental activity: water – D_e = 2; kerosene – D_e = 2.1; acetone – D_e = 2.46; benzene – D_e = 2.56. Thus for water the relaxation occurs along structural surfaces whereas for benzene it is provided by the fractal objects that are close to the volumetric ones.

3. Conclusions

Thus the presented theoretical and experimental studies confirm the urgency of the task – the search for new kinetic models which give a noncontradictory evaluation of the physical phenomena and reveal their nature. In order to solve this task the kinetic model developed using the concepts of the theory of topological dynamics was used which allowed giving a qualitatively new understanding of the relaxation processes.

References

[1] Voronov P V, Bobryshev A N, Zharin D E, Lakhno A V and Shafigullin L N 2011 Rheology of viscous composite systems with heterogeneous structure *Vesntik Kazan state technical university named after A N Tupolev* **3** 13-17
[2] Voronov P V, Bobryshev A N, Zharin D E, Lakhno A V and Simonova K V 2011 Model of transitional creep of polymer composites *Vesntik Kazan state technical university named after A N Tupolev* **1** 27-33
[3] Mashnin N N 1975 *Applied theory of plasticity and creep* (Moscow: Mashinostroenie) 398
[4] Bobryshev A N, Erofeev V T and Kozomazov V N 2012 *Physics and synergetics of dispersion-disordered condensed composite systems* (St. Petersburg: Nauka) 473
[5] Bartenev G M 1984 *Strength and fracture mechanism of polymers* (Moskow: Khimiya) 279
[6] Postnikov V S 1978 *Physics and chemistry of solid state* (Moskow: Metallurgiya) 540
[7] Shklovsky B I and Efros A L 1979 *Electronic properties of doped semiconductors* (Moskow: Nauka) 279

IOP Conf. Series: Materials Science and Engineering **86** (2015) 012002 doi:10.1088/1757-899X/86/1/012002

High voltage high repetition rate pulse using Marx topology

A.Hakki , N.Kashapov
Kazan (Volga-Region) Federal University, Tatarstan, Russian Federation

E-mail: hakki.adnan@gmail.com

Abstract. The paper describes Marx topology using MOSFET transistors. Marx circuit with 10 stages has been done, to obtain pulses about 5.5KV amplitude, and the width of the pulses was about 30μsec with a high repetition rate (PPS > 100), $V_{dc} = 535\text{VDC}$ is the input voltage for supplying the Marx circuit. Two Ferrite ring core transformers were used to control the MOSFET transistors of the Marx circuit (the first transformer to control the charging MOSFET transistors, the second transformer to control the discharging MOSFET transistors).

Key words. Marx Topology, MOSFET transistors, High Voltage Pulse

1. Introduction

The conventional Marx is not suitable for most of the applications because it needs long charging time (the charging current flows through the charging resistors), Low efficiency, Low repetition rate, few output voltage appearance in charging period, "turn off" is impossible because of using the spark gap switches, short life time of the spark gap switches [1-5].

2. Experimental Setup

Fig.1 demonstrates the Marx circuit with two stages (in our case we have 10 stages). The MOSFET transistors (Tch1-Tch10) constitute a charging group, and the MOSFET transistors (Tdisch1-Tdisch10) constitute the discharging group. (C1-C10) are the storage capacitors.

IOP Conf. Series: Materials Science and Engineering **86** (2015) 012002 doi:10.1088/1757-899X/86/1/012002

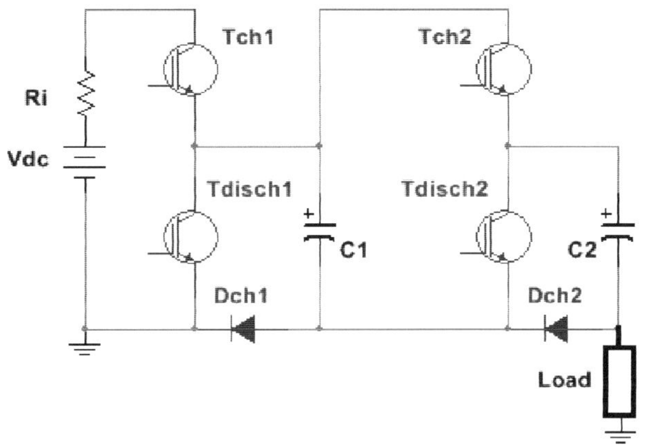

Figure 1. Marx circuit with two stages

Fig.2 demonstrates the equivalent charging circuit with two stages (in our circuit we have 10 stages). In this case of the MOSFET transistors (Tch1-Tch10) are in "turn on" position, and the MOSFET transistors (Tdisch1-Tdisch10) are in "turn off" position. All capacitors (C1- C10) are charged via diodes (Dch1-Dch10). For example, the capacitor (C1) is charged via the diode Dch1.
The input DC voltage is given by:

$$V_{dc} = V_{IN} \approx 1.41V_{p-p} \approx 535\text{V}$$

Where $V_{p-p} = 380V$ is a phase to phase voltage.

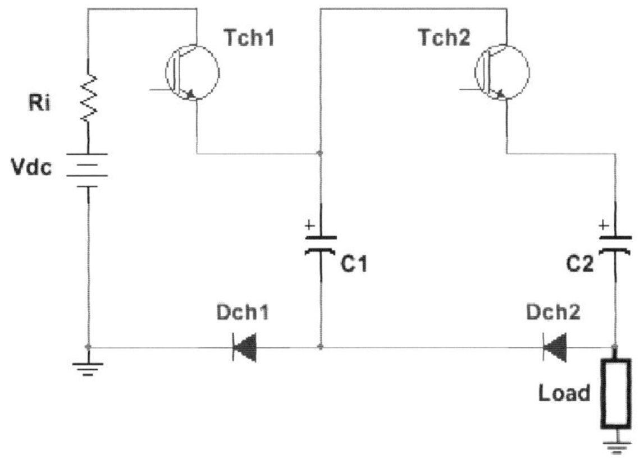

Figure 2. Equivalent charging circuit with two stages

Fig.3 demonstrates the equivalent discharging circuit with two stages (in our circuit we have 10 stages). In this case the MOSFET transistors (Tdisch1-Tdisch10) are in "turn on" position and the

MOSFET transistors (Tch1-Tch10) are in "turn off" position. All capacitors (C1- C10) are discharged via the load resistance.

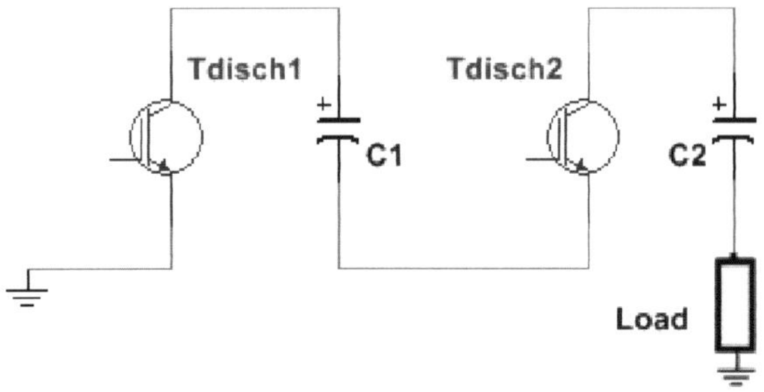

Figure 3. Equivalent discharging circuit with two stages

Note that the charging voltage of each capacitor is V_{dc}.
Fig.4 demonstrates the all signals of Marx circuit.
The output voltage is given by:

$$V_0 \approx \text{n}.V_{dc} \approx 5.5\text{KV}$$

Where n: Number of stages (in our case $n = 10$)
t_c: off time (charging time).
t_d: on time (discharging time).

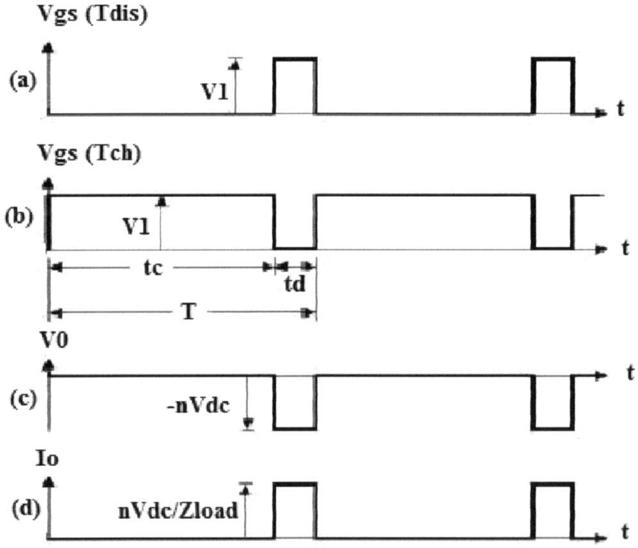

Figure 4. Signals of Marx circuit
(a) Control signals of the transistors T_{disch1} - $T_{disch10}$, (b) Control signals of the transistors T_{ch1} - T_{ch10}, (c) Output voltage pulse, (d) Output current pulse

Fig.5 demonstrates the control circuit of the MOSFET transistors (charging). The input control signal U1 of the circuit originates from the control unit.
The transformer T1 generates the pulses Vgs1-Vgs10 (as shown in Fig.4-b) in order to control the MOSFET transistors (Tch1-Tch10). It means that all transistors (Tch1-Tch10) are in same position.
During the charging period, all the transistors (Tch1-Tch10) are in "turn on" position. On the other hand, all the transistors (Tdisch1-Tdisch10) are in "turn off" position as shown in Fig.2.

The storage energy in the capacitors given by [6]:

$$E_C = \frac{n}{2} C_i {V_{dc}}^2$$

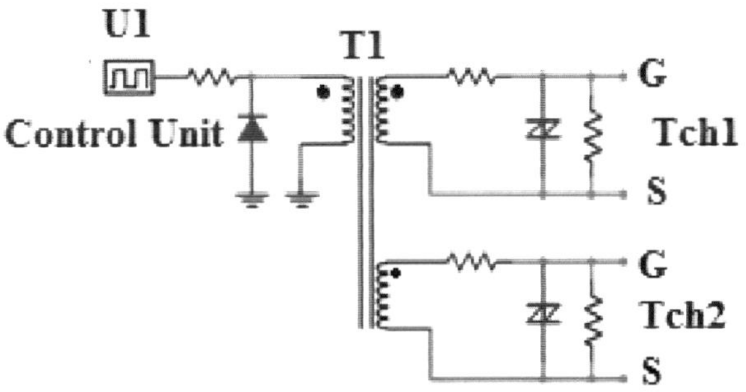

Figure 5. A control circuit (two stages) of the MOSFET transistors (Charging)

Fig.6 demonstrates the control circuit of the MOSFET transistors (discharging), where the input control signal U2 of this circuit originates from the control unit.
The transformer T2 generates the pulses Vgs1-Vgs10 (as shown in Fig.4-a) to control the MOSFET transistors (Tdisch1-Tdisch10). It means that all transistors (Tdisch1-Tdisch10) are in same position.
During the discharging period all the transistors (Tdisch1-Tdisch10) are in "turn on" position and all the transistors (Tch1-Tch10) are in "turn off" position. This is shown in Fig.3.
The pulse energy during the discharging period t_d is given by [7]:

$$E_{Pulse} = nV_{dc} i_o t_d$$

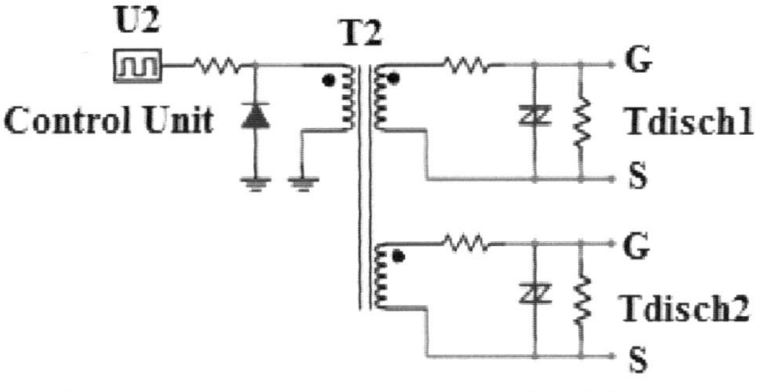

Figure 6. A control circuit (two stages) of the MOSFET transistors (Discharging)

3. Conclusion

Marx circuit with 10 stages was made, in order to obtain pulses that have the amplitude of about 5.5KV, and the width of the pulses is about 30μsec with high repetition rate (PPS > 100). V_{dc} = 535VDC is the input voltage used for supplying the Marx circuit. Two Ferrite ring core transformers were used to control the MOSFET transistors. The circuit shown in Fig.1 had a very short charging time, high efficiency, and a high repetition rate. These pulses can be used in many applications such as a plasma ignition, flash lamp trigger.

The work is performed according to the Russian Government Program of Competitive Growth of Kazan Federal University.

References

[1] T. Iwashita, T. Adachi, K. Takayama, T. Arai, Y. Arakida, M. Hashimoto, E. Kadokura, M. Kawai, T. Kawakubo, Tomio Kubo, K. Koyama, H.Nakanishi, K. Okazaki, K. Okamura, H. Someya, A. Takagi, A. Tokuchi, K. W. Leo, and M. Wake, "KEK Digital Accelerator", to be published in Phys. Rev. ST-AB 14, 071301 (2011).
[2] S.C.Haydon, The physics of lightning Proceedings Symposium on Non-conventional Lightning Protection, Paper No, 1,Inst.Eng, Aust, Sydney, October, 1986.
[3] M.Giesselmann, B.Palmer, A.Neuber, Texas Tech University, high voltage impulse generator using HV-IGBTs, Center for Pulsed Power & Power Electronics Dept. of Electrical & Computer Engineering. USA
[4] Kornienko, E.E., Lapushkina, E.J., Kuzmin, V.I., Vaschenko, S.P., Gulyaev, I.P., Kartaev, E.V., Sergachev, D.S., Kashapov, N., Sharifullin, S., Fayrushin, I. 2014 J. Phys.: Conf. Series **567** 012010
[5] Israfilov, Z.Kh., Kashapov, N.F. 1991. Journal of Engineering Physics 60 (3), pp. 364-368
[6] M.Uman, The lightning discharge international geophysics series . Volume 39.
[7] T.Adachi, T.Arai, K.W.Leo, K.Takayama, A.Tokuchi, "A Solid-state Marx Generator Driven Einzel Lens Chopper", Rev. of Sci. Inst. 82, 083305 (2011).

IOP Conf. Series: Materials Science and Engineering **86** (2015) 012003 doi:10.1088/1757-899X/86/1/012003

Methodology development for the sustainability process assessment of sheet metal forming of complex-shaped products

D L Pankratov, L R Kashapova

Kazan University, Naberezhnye Chelny Institute
423810, Naberezhnye Chelny, Mira av., 68/19, Russia.

E-mail: pishite.lile@gmail.com

Abstract. A methodology was developed for automated assessment of the reliability of the process of sheet metal forming process to reduce the defects in complex components manufacture. The article identifies the range of allowable values of the stamp parameters to obtain defect-free punching of spars trucks.

1. Introduction

Stamping is one of the most popular and effective methods of production. Stamping methods in most cases achieve the necessary quality parts (precision linear dimensions and relative position of surfaces, surface roughness), high performance and low cost. [1]

However, during products stamping, especially of complex forms, defects often occur such as dents; folds, tear sheet or cracks. It is important to note that the occurrence of defects is of a periodic nature, it is caused, first, by the difference in the quality of the source material, depending on its supply and, secondly, the operation of the tool in the limiting values of geometrical parameters of its deforming parts. Today, it is important to have a way to eliminate such defects, because the production of high quality parts is one of the main tasks of the leading companies in the world.

2. Basic part

After analyzing the causes of defects, factors of formability have been identified [2, 3]. These factors are indicators of the reliability of the sheet metal forming process, its content and structure, they were classified and presented in the profile view in [4]. The profile presents four groups of the reliability factors: material properties, stamping technology, geometric parameters of the tool and equipment parameters, which allow to combine different sizes of techno-economic parameters of the product in one dimensionless figure – integral coefficient of reliability and objectively evaluate and compare the technological processes. [5]

In [4] an algorithm to evaluate the reliability of the technological process of sheet metal forming also presents.

Reliability is understood as a property of the object to save time and within the established values of all parameters characterizing the ability to perform a required function under given conditions and conditions of use. Reliability is a complex property. [6]

As an example, spars trucks bends forming in the locations of the rear wheels was considered. This site is defective, because cracks and folds in the so-called elongation zone and the zone of coagulation take place. These zones are shown in figure 1.

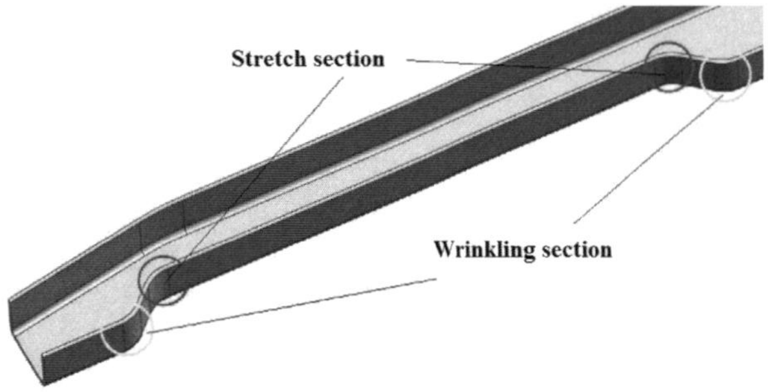

Figure 1. Defective zones in the elongation zone

Reliability factors "Material Properties" and "Hardware Settings" in this case are considered in the form of certain "constants", since they are the source data to the technological process and, accordingly, determine cause of defects. Therefore, for instance, these parameters are only possible to issue recommendations: apply additional material processing, conduct unscheduled repairs, equipment upgrades, measures to increase tool life, etc. [7 - 9]

Reliability factors "Punching Technology" and "Geometrical parameters of the tool" are controlled, because the required coefficient of friction can be achieved by applying a lubricant, and the necessary clearance between the workpiece and the tool can be achieved by changing the shape and dimensions of transition edges of working parts of the stamp. The list of these parameters is shown in table 1.

Thus, the profile of the reliability indices for the considered case will look as shown in figure 2.

Formability factors are the following: irrational geometry of the stamp deforming elements, damaged edges of the initial workpiece, temperature and the coefficient of friction between the surface of the workpiece and the working parts of the tool.

The source material is subjected to incoming control, and the temperature is maintained according to the manufacturing process. Therefore, to obtain defect-free stamping, it is necessary to consider the influence of geometrical parameters of the working parts of the stamp (in this case, it is the radius of the transition edge of the matrix) and to investigate the influence of contact friction in sheet metal forming process.

For simplicity of calculations, a part of the deformable region was taken (figure 3).

According to the simulation, graphic charts of values thinning under varying parameters stamp were built: $L_R = f(R) -$ distribution curve values thinning φ depending on the radius of rounding of the matrix R edges; $L_{[\varphi]} = 35\% -$ curve of valid values thinning φ for steel ([φ]=35%); $L_1 = 16$, $L_1 = 48-$ curves, expressing the condition of boundedness of the corner radii of the matrix edges ($16 \le R \le 48$ mm). [10]

According to the graph, a range of allowable values of the parameters of the stamp was identified to obtain defect-free stamping (in figure 4 it is a shaded area), i.e. the radius of the rounded edges of the matrix must be in the range of 27.5°≤R≤48°.

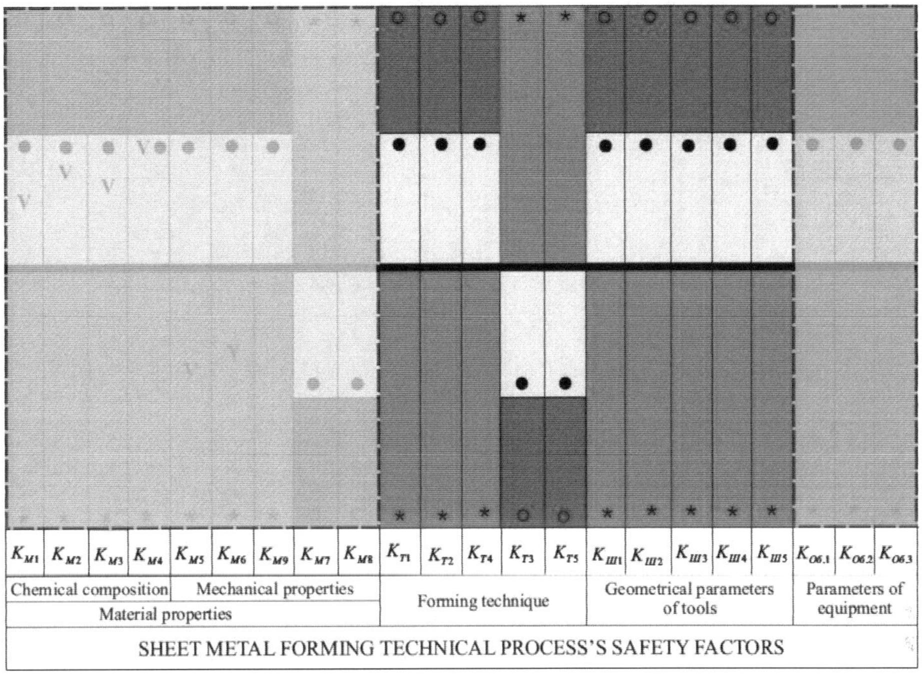

Figure 2. Reliability profile of the sheet metal forming technological process

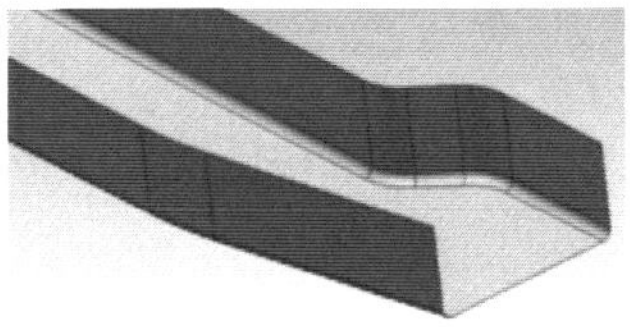

Figure 3 – Plot of the spar for calculations in LS-DYNA

The effect of the coefficient of friction between the surface of the workpiece and the working parts of the tool was further investigated. The simulation was conducted in three phases (three groups): with a constant coefficient of friction (Group №1); with a variable coefficient of friction in the system Procurement-Matrix and a constant coefficient in the system Plug-Procurement (Group No. 2); with a variable coefficient of friction in the system Plug-Procurement and the constant coefficient in the system of Procurement-Matrix (Group No. 3).

The curves are presented in figure 4.

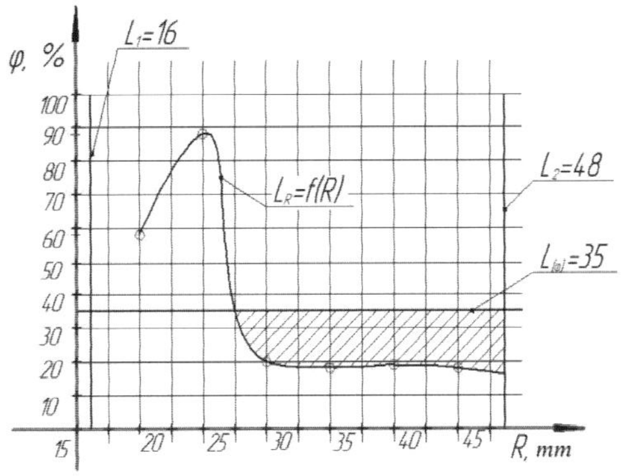

Figure 4 – Graphs of thinning values dependency from the matrix edge fillet radius

Table 1. Safety factors in group "Stamping technology" and "Geometric parameters of the tool".

Safety factors	Name title of safety factors
K_{T1}	the coefficient of friction between the surface of the workpiece and the working parts of the tool
K_{T2}	the coefficient of permissible deviations in thickness of the original sheet
K_{T3}	the coefficient of thermal regime
K_{T4}	the coefficient of permissible deviations of the size of the original sheet
K_{T5}	the limiting drawing ratio
$K_{Ш1}$	the coefficient reflecting the hardness of the material of the wetted parts of the tool
$K_{Ш2}$	the coefficient reflecting the value of the roughness of the surfaces of the working parts of the tool
$K_{Ш3}$	coefficient taking into account the wear of the working parts of the tool
$K_{Ш4}$	coefficient taking into account the misalignment of the upper and lower parts of the stamp
$K_{Ш5}$	coefficient taking into account the uneven gap between the working parts of the stamp

Due to the obtained data graphs of the metal thinning, depending on the friction coefficient, were constructed (figure 5). According to the obtained data, the conclusion can be made concerning the dependence of the measure of friction, i.e., the greater the friction, the greater the thinning.

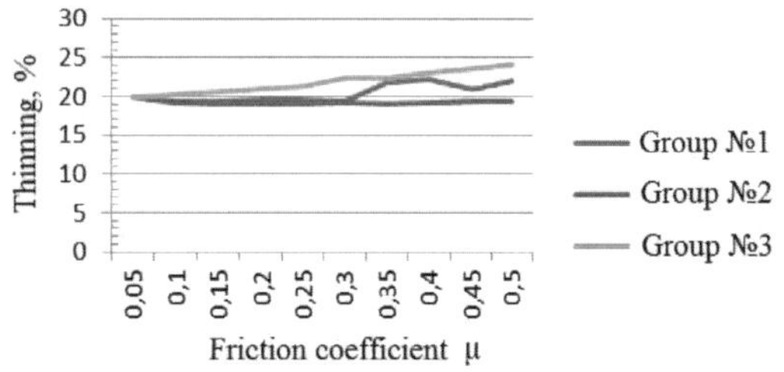

Figure 5. Graph of dependence between thinning and friction coefficient

If we consider modeling group separately, we can see that Group No. 2 practically does not change its characteristics, regardless of the friction coefficient. It can lead to the conclusion, that the influence of the friction coefficient on its reliability in this process is insignificant. Group No. 3 is characterized by a gradual increase in thinning depending on the increase of the friction coefficient. Group No. 1 is characterized by a sharp increase in thinning when exceeding the coefficient of friction value of 0.35. Thus, the practicability is in side members stamping when the coefficient of friction is less than 0,35.

3. Conclusions

Thus, it is possible to obtain defect-free stamping by changing the geometric parameters of the working parts of the stamp or by controlling the value of the friction coefficient. The work is aimed at solving urgent practical tasks – implementation of defect-free punching of complex shape sheet parts with a minimum expenditure of manpower and material resources for production preparation.

References

[1] Kovalev V.G. Technology of sheet metal forming [Tehnologiya listovoy shtampovki], M.: KNORUS, 2010, 224p.
[2] Elenev, S.A. Cold stamping [Holodnaya shtampovka], M.: Vyisshaya shkola, 1988. – 271p.

[3] Semenov, E.I. Forging and stamping. The guide. Vol. 4. Stamping. [Kovka i shtampovka. Spravochnik v 4 t. T.4 Listovaya shtampovka], M.: Mashinostroenie, 1987. – 544p.
[4] GOST 27.002 – 89. Reliability in engineering. The basic concepts. Terms and definitions [Nadezhnost v tehnike. Osnovnyie ponyatiya. Terminyi i opredeleniya], 64p.
[5] UDK 658.56 Differential method of assessing the quality of hire [Raznostnyiy metod otsenki kachestva avtomobiley]. H.A. Fashiev, I.D. Valeev // Avtomobilnaya promyishlennost, 2007, #11, pp.3-7.
[6] Kashapova L.R., Pankratov D.L., Shibakov V.G. Methodology for automated assessment of the reliability of the sheet metal forming process at the stage of its design [Metodika avtomatizirovannoy otsenki nadezhnosti tehnologicheskogo protsessa listovoy shtampovki na etape ego proektirovaniya] // Fundamentalnyie issledovaniya, # 8 (chast 7) 2014, pp.1533-1538.
[7] Zuev V.M. Heat treatment of metals: textbook for technical schools [Termicheskaya obrabotka metallov: ucheb. dlya tehnicheskih uchilisch], 2-e izd., pererab. i dop. M.: Vyissh. shk., 1981. – 296p.
[8] Romanovskiy V.P. Handbook of cold stamping [Spravochnik po holodnoy shtampovke], L.: Mashinostroenie, 1979. – 782p.
[9] Goryaynov V.I. Cold stamping equipment and its commissioning. [Holodnoshtampovochnoe oborudovanie i ego naladka. Ucheb. posobie dlya SPTU], M.: Vyissh. shk., 1988. – 256p.
[10] Rudman L.I., Marchenko V.L., Zaychuk A.I. i dr. Stamps designer handbook. Stamping [Spravochnik konstruktora shtampov. Listovaya shtampovka], M.: Mashinostroenie, 1988, 496p.

Technology of manufacturing of multiple core stamp tools by plastic deformation

D L Pankratov, V G Shibakov, A M Valiev, R F Valieva

Kazan Volga Federal University, Naberezhnye Chelny Institute
423810, Naberezhnye Chelny, Mira av., 68/19, Russia.

E-mail: pankratovdl@gmail.com

Abstract: This article describes a new method for manufacturing of multiple core stamp tools by plastic deformation. The technique of geometric parameters of connected elements of multiple core tools designing necessary for the production of the proposed technology is developed.
Key words: punch, stamping tools, blank, assembling by plastic deformation

1. Introduction

One of the main factors that significantly affects the durability of the tool for punching holes in the stamping manufacture is a complex of the mechanical properties of the material from which the working elements of the tool are made [1; 2; 3]. For example, the resistantance of punches made from U10A steel is 40-80 thousand strikes and for the punches made from X12M steel - 150-200 thousand ones [4].

Thus the manufacturing costs of steel punches X12M are 2.5-2.7 times higher than that one of similar design and purpose made of steel U10A. Reducing tooling costs by saving expensive alloy steels is an important task of tool production. In some cases this problem can be solved by applying a compound tool, wherein the most loaded part of the working tool is made of tool steel and other construction elements from a cheap carbon one.

2. Basic part

For manufacturing of such class tool is offered a new way, based on the open punching of a blank.

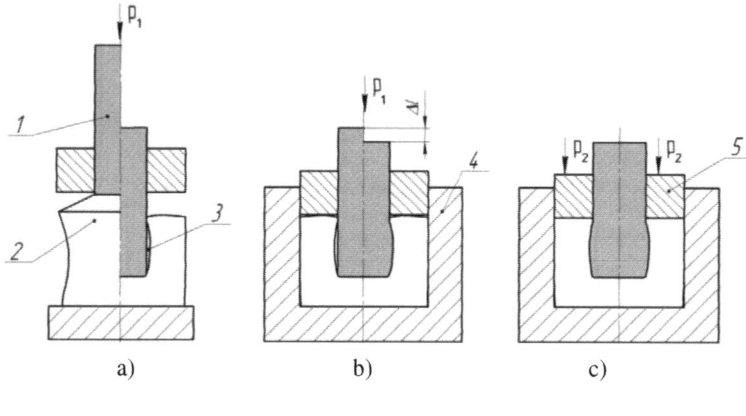

Figure 1. A method of manufacturing of a bimetallic core tool by punching.

The proposed technology of the compound punching tool by punching consists of four stages. In the first stage by the penetration of core 1 is carried out an open broach of body blank 2 (Figure 1a) heated up to forging temperature, while in the process of penetration owing to metal displacement features in the open punching between the cylindrical surface of the core and the side surface of the body blank cavity is formed an annular gap 3. The contoured shape of the body blank is selected taking into account the non-uniformity of deformation, hot upset in a height at the open punching. This shape provides a cylindrical shape of the body blank after open punching, i.e with minimum barrel shape of the side surface without sink mark, but forming an annular gap θ_3.

At the second stage after the localized heating of the assembled blank a hot upset of the rod part by an amount Δl (Fig. 1b) is made. Thus there is a prefabricated compound in the matrix 4, due to the filling of the annular gap by a rod material.

At the final stage the final assembly of permanent connection by hot upset of body blank by an annular tool 5 in order to create the contact pressure sufficient to forge welding on the bonding surfaces (Fig. 1v). Theoretical fundamentals of the constituent permanent joints obtaining composite permanent joints by a plastic deformation are described in details in the works [5; 6; 7].

To implement this technology of precast rod stamp tools by a plastic deformation is necessary to develop a technique of contour blank designing. For this purpose, the process of form changing of cylindrical blanks in the open punching is studied.

Figure 2 shows the calculation scheme for determining the form changing parameters in the open punching with the diameter $d = 0{,}3 \div 0{,}7D_0$. The choice of this range is due to the fact that when $d < 0{,}3D_0$ the deformation of the blank is local, since the process of punching begins to move into the process of pressing the punch in the half-space that goes beyond the general case of deformation at the open punching, and if the size of the punch $d = 0{,}7 \div 1{,}0D_0$ due to the small immersion of the punch it does not provide a blind hole in the blank [8], which excludes the possibility of obtaining a permanent connection.

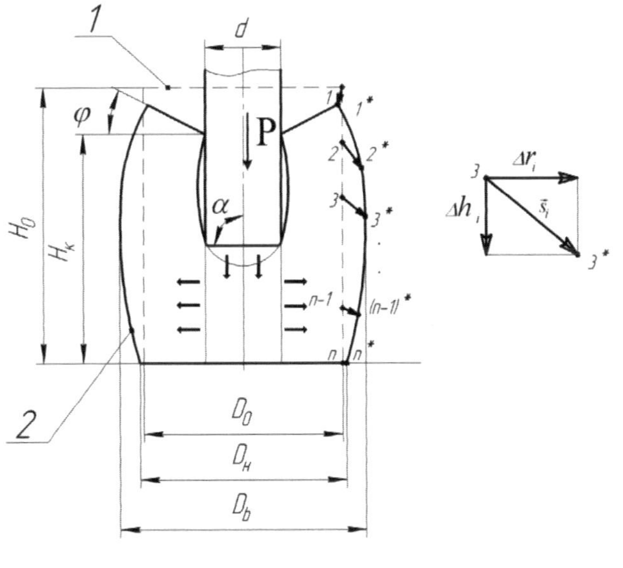

Figure 2. Analysis of forming a cylindrical blank at the open punching

Basic regularities of the form changing of a cylindrical blank at the open punching are considered in the works [9; 10]. Blank dimensions before punching are characterized by height H_0 and the diameter D_0 of the blank (contour 1). As a result of punching the blank shrinks, and its height becomes H_K. The preform acquires the shape of a barrel with a barrel diameter D_b (contour 2). By figures 1, 2,… n are shown the initial situations of the equidistant points on the lateral surface, wherein n – number of points. By the figures 1*, 2*,… n* are shown the final positions of the same points after punching. Moving from the initial position is characterized by a vector $\vec{s}_i(\Delta r_i, \Delta h_i)$, wherein Δr_i and Δh_i– amount of displacement i – points in the radial and axial directions, respectively.

Area of the constrained deformation 3 moves downward, pushing the metal from the deformation source 4 in the radial direction, moving apart the annular region 5. The bending of the sample side surface leads to the formation of a gap between the punch and punching element whose volume is $V_з$. The volume of the annular gap in the punching can be characterized by the following relative value

$$\theta_з = \frac{V_з}{V_{в.п.}} 100\% , \tag{1}$$

wherein $V_з$ – the volume of the annular gap, мм3; $V_{в.п.}$ – the volume of the embedded portion of the cylindrical punch, mm3.

The contour of the upper end surface rotates by an angle φ, that forms a sink marks.

On the basis of experimental studies are obtained the mathematical models of size dependence of the final form changing of the blank Δr_i, Δh_i, D_b, $D_н$, $H_к$, φ, $\theta_з$ on a diameter d/D_0, blank height H_0/D_0, angle at the base of the punch α. Accepted values of the varied factors: the diameter of the punch d/D_0 = 0.3÷0.7; blank height H_0/D_0 = 0,5÷1,5. (for large values H_0/D_0 it is difficult to ensure the stability of disposable punch); angle at the base of the punch α = 45° ÷ 90°. The depth of the punch penetration - $0{,}8H_0$ (greater depth of punching leads to a removable jumper and is mainly used for punching of a through hole in the blank).

Based on these studies a method of designing the geometric parameters of the contoured blank according to the drawing of the final compound product is developed. The first step is the calculation of the dimensions of D_0 and H_0 cylindrical blank 3, the annular gap value $\theta_з$, taper rod angle α and length of the immersed part of the rod $l_{вн.}$.

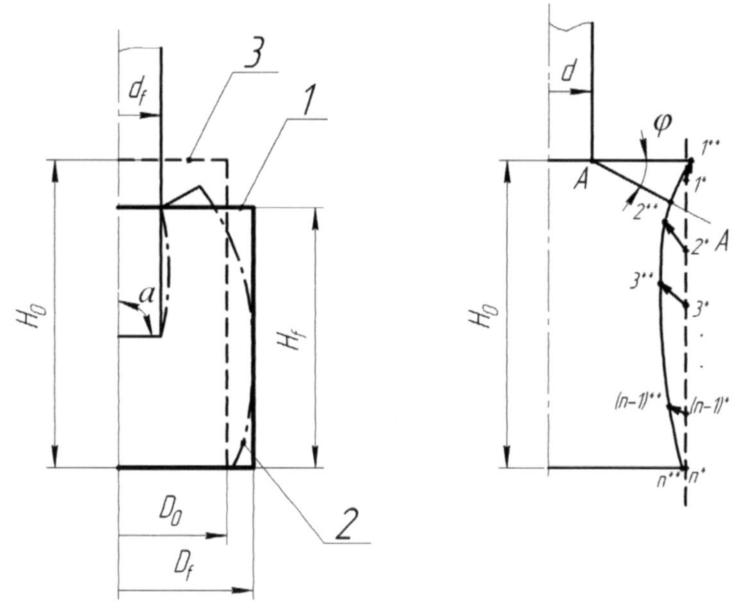

a) b)

Figure 3. Designing of the contoured blank

The contour of the given cylindrical blank 2 after punching has to fit into the contour of the final compound tool 1 with the dimensions d_f, D_f и H_f (fig. 3, a).

IOP Conf. Series: Materials Science and Engineering **86** (2015) 012004 doi:10.1088/1757-899X/86/1/012004

After determining the dimensions of the cylindrical blank point displacement values of its side surface are calculated in the radial Δr_i and axial Δh_i directions and the sink marks angle of the upper end surface φ. The obtained values are used in the construction of countered blank contour (see. Fig. 3b).

The possibility of implementing of the proposed technology to manufacture precast rod stamp tools by plastic deformation is confirmed as by the results of simulation in the finite element modeling program complex as by manufacturing of experimental samples.

References

[1] Reference book of stamp designer: Stamping /, ed. L.I. Rudman., M .:
Machinery, 1988. - 496 p .: ill.

[2] Averkiev Y.A., A. Y. Averkiev Cold forging technology: Textbook for universities
the field of "Machines and technology of metal forming" and "Metal Forming.", M .: Engineering, 1989. - 304 p .: ill.

[3] The dies for stamping. Calculation and design. Reference book ed. V.V. Belov and G.I. Khesin, L: 1992, 295 p.

[4] V. Romanovsky Reference book of cold forming. - 6 th ed., Rev. and add. - L .: Mechanical engineering. Leningrad. Div., 1979. - 520 p., ill.

[5] Bases of pressure welding. Gelman A.S., M., "Engineering", 1970, 312 p.

[6] The technology of composite metals: Tutorial. Kobelev A.G., Potapov I.N., Kuznetsov E.V. - M .: Metallurgy, 1991, 248 p.

[7] A.A. Chularis Pressure welding technology / A.A. Chularis, D.V. Rogozin. - Rostov on / D: Phoenix, 2006. - 221, [1] p .: with ill. - (Higher Education)

[8] J.M. Okhrimenko and Tyurin V.A. The theory of the forging process. Textbooks for higher education. M., "High School ", 1977. 295 p., ill.

[9] Berezkin V.G. Form changing in metal forming M .. "Engineering", 1973, 152 p.

[10] Storojev M.V., Popov E.A. Theory of Metal Forming. Textbook for High school. 4th ed., Revised. and add. M., "Engineering", 1977. 423 p., with ill.

The appearance of shock waves in the plasma electrolytic processing

D.G. Denisov[1], N.F. Kashapov[1], R.N. Kashapov[2]
[1] Kazan Federal University, 18 Kremlyovskaya St
[2] Kazan Physical -Technical Institute, Sibirsky tract 10/7

E-mail: kashramil.88@mail.ru

Abstract. In this paper, we have been tasked to investigate the transitional regime, to identify the main physical phenomena and give them a qualitative explanation. As a result of experimental researches, have been proposed and discussed possible mechanisms for generating shock waves in the transition region.

Introduction

Discharges in liquids for a long time are interesting to researchers, due to the presence of a wide range of physical and chemical phenomena. One of the kinds of these systems is discharges with liquid electrodes when one of the electrodes (anode or cathode) is an electrolyte solution. Well enough investigated the discharge in a stationary mode at voltages higher than 200 V while the metal electrode immersed and situated above the surface [1]. These systems have been used in the processes of thermal effects [2], in the processes of plasma-chemical treatment processes [3], in the formation of protective coatings [4], in the process of cleaning fluid from toxic pollution. However, it remains unclear mechanism of transition from electrolysis in the stationary mode burning of the discharge [5] and did not explain the reasons for the physical and chemical phenomena that occur at the same time. In this paper, we have been tasked to investigate the transitional regime, to identify the main physical phenomena and give them a qualitative explanation.

Experimental

For research the transition regime was used experimental setup, shown in Fig. 1. It consists of the electrolytic cell 1, power supply 2, test equipment 3, shunt 4, the electrode system 5, thermometer 6, high-speed video cameras 7 and ammeter 8.

System electrical power is a DC power supply with continuously variable voltage [6]. The power supply consists of a diode bridge and laboratory autotransformer adjustment type 1M with a voltage range from 1 to 240 V, 50 Hz. Used the work in two modes, the first - when the applied voltage obtained after full-wave rectification, and the second - with the additional use of a capacitive filter capacitance C = 1560 uF.

Research of the transition regime conducted on the anode made of graphite and steel grades of Russian steel #3, 12X18H10T. To prevent resizing dielectric protective cover, which leaves open only a strictly defined part of the electrode, was worn on the area of the anode. The area of the anode was 0.65 cm2. As cathode was used copper plate, an area of 35 cm2. The electrolytic bath is filled with researched electrolytes in necessary concentration and composition. Aqueous solutions (NaCl, Na_2CO_3, NaOH) were used as the working fluid.

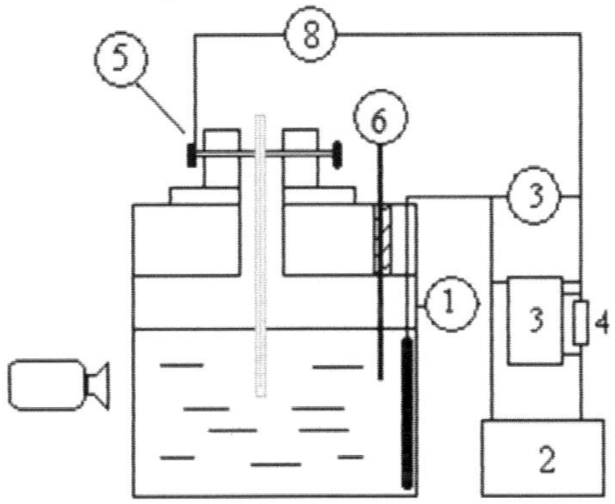

Fig. 1 Functional diagram of the experimental setup

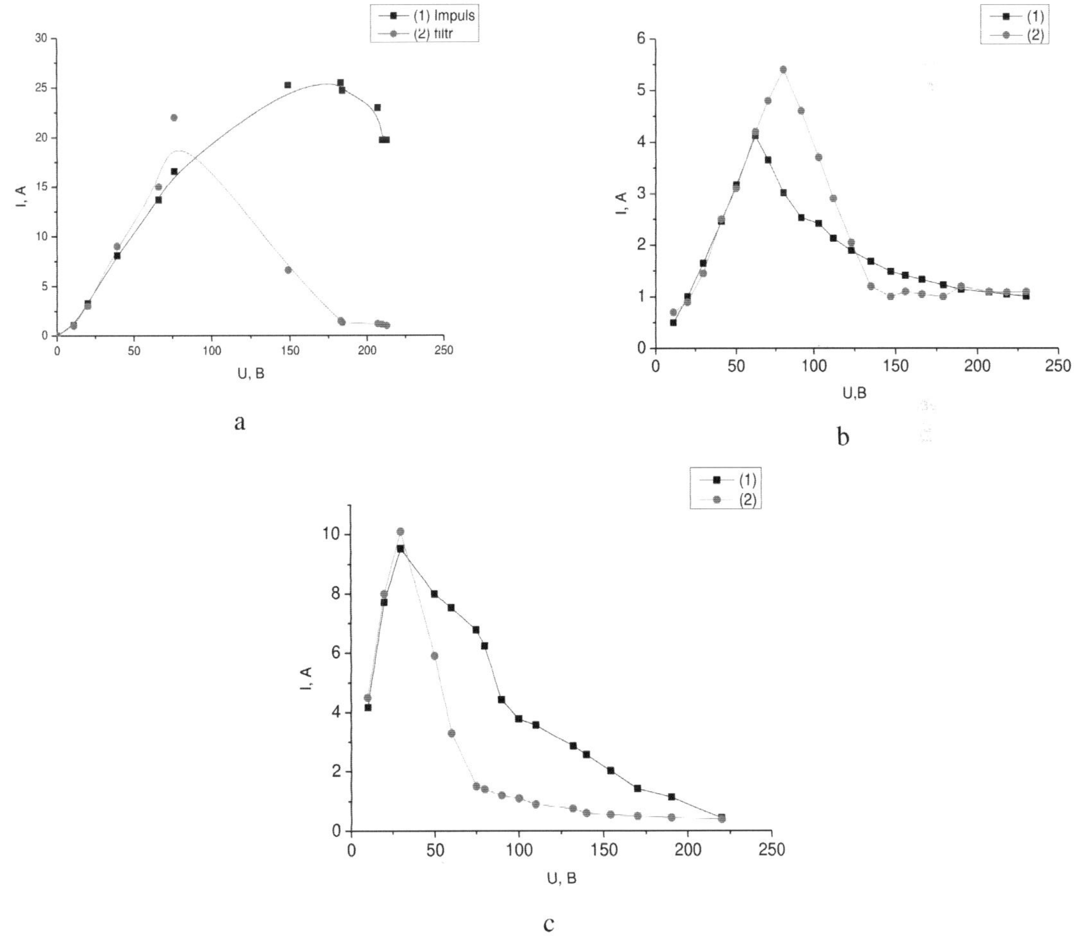

Fig. 2. CVC of plasma electrolytic process. a - 5% aqueous solution of NaCl; b - 5% aqueous solution of Na_2CO_3; c - 5% aqueous solution of NaOH. (1) - pulsed (2) - smoothed.

The main operating parameters of the system are: the voltage on the electrodes, the magnitude of the discharge current, the current density at the electrodes, the electrolyte temperature and the temperature of the anode. Changing the shape of voltage and current at the time of discharge ignition was determined using an oscilloscope FLUKE 190-062 SCOPEMETER 2CH 60MHz 625 MS / s, time base varies from 5 ns to 60 s. The relative measurement error did not exceed 0.025%. Measure the voltage at the anode was carried out using a digital universal measuring device APPA-207, the relative error of measurement is 0.5%. With a digital multimeter APPA 305 temperature measurement was carried out of the electrolytic cathode in using the chromel- copel thermocouple. Measuring the temperature of the anode was carried out by thermometer Fluke 62 Max. Research of fast processes in the transition region of plasma electrolytic processing was carried out by using high-speed video camera Fastec HiSpec 4. The shooting speed was regulated, at a maximum resolution of 1696 * 1710 amounted to 523 frames per second at 320 * 240 - 14 781 frames per second.

Processing of the results was carried out taking into account the measurement uncertainty.

Before the research of the transition region current-voltage characteristics of plasma electrolytic process in a pulsed mode and smooth were obtained, presented in Fig. 2.

When using a capacitive filter established that the transition occurs during 20 millisecond. From the end of the electrode gas bubble growing, which gradually rises up and covers the entire surface.

Also, there is a strong 90% drop in current and voltage increase by an average of 15 V. Of interest is the transition in the pulse mode. When using a full-wave voltage of generation shock waves were observed. Current-voltage characteristics of full-wave mode can be divided into three sections. The first section from 0 to 70 V. This area corresponds to the course of only electrochemical reactions. The second region 70 V to 180 V – it is the burning of separate microdischarges at the anode surface. Precisely in this range, shock waves are generated. In the third area of 180 V and above there is a glow of vapor-gas layer around the electrode without the occurrence of certain microdischarges. The process discharge burning occurs intermittently, with each new pulse, appearing again and again. Nature of the electrolyte is generally not effects the picture of occurring physical processes. The difference lies in the strength of the current and the borders of the transition from the electrolysis mode of burning anode discharges. The use of a graphite electrode, also not show value changes.

Consider a series of photographs presented in Fig. 3, (one frame corresponds to one millisecond) full-wave mode, obtained at the average voltage of 80 V and a current of 20 A. It can be divided into three stages. First - electrochemical reaction, frame №1. Further, when the ignition voltage on the anode surface is initiated of burning anode microdischarges - this is the second stage, which lasts 3 ms.(2, 3 and 4 frames). The third stage - is the formation and growth of gas bubbles (5, 6 frame) centered on the place where previously burned microdischarges. Thus, the period of the current 10 ms consists of the area of electrochemical reactions lasting 4 ms, of burning gas microdischarges - 3 ms, the formation and growth of the gas bubbles - 3 ms. But it should be noted that the development and exit gas bubble to the surface of electrolyte may take up to 20 ms.

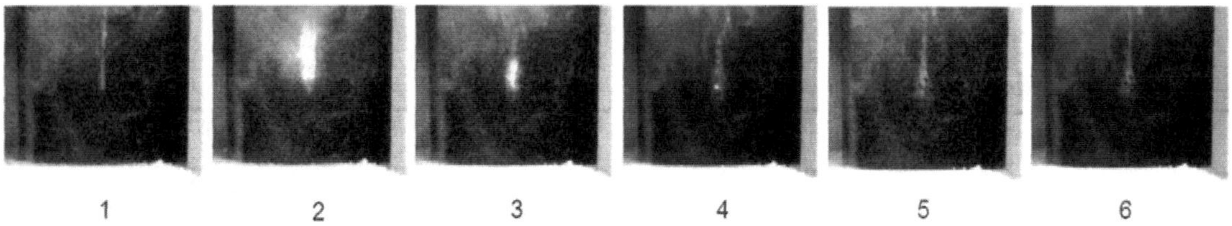

Fig. 3. Series of high-speed video frames.

Let's look at the possible mechanisms to generate shock waves in the transition area. The first of the options is the presence of microbubbles, which are polarized in an electric field and under the periodic influence of 100 Hz are beginning to waver. Characteristic feature is presence of sound, with no visible light radiation. Further may occur two possible options: the collapse of the bubble and the formation of a shock wave (similar to cavitation) or the breakdown of a gas bubble, a sharp drop in pressure in the bubble after quenching of the discharge and the collapse of the bubble, which also lead to the emergence of the shock wave.

The second possible mechanism is that the calefaction of the discharge channel occurs inside the cavity of the leader, in the initial stage of calefaction does not occur over the entire cross section of cavity leader channel. Thus, at the initial moment, we are seeing the formation of a leader in the liquid and its germination in environment at a small distance, and then localized calefaction in the gas discharge

References

[1] Allagui, A., Elwakil, A. On the N-shaped conductance and hysteresis behavior of contact glow discharge electrolysis. Electrochimica Acta Volume 168, 20 June 2015, Pages 173-177
[2] Kashapov L, Kashapov N, Kashapov R 2013 *J. Phys.: Conf. Series* **479** (1), 012003
[3] Kashapov L, Kashapov N, Kashapov R 2013 *J. Phys.: Conf. Series* **479** (1), 012011
[4] Kashapov L, Kashapov N, Kashapov R 2013 *J. Phys.: Conf. Series* **479** (1), 012005
[5] Kashapov R.N. Investigation of plasma-electrolytic process of forming micro-relief surface of metals. The dissertation for the degree of candidate of technical sciences. 2012.
[6] Fadeev S A, Kashapov N F and Larionov V M 2014 *IOP Conf. Ser.: Mater. Sci. Eng.* **69** 012006

IOP Conf. Series: Materials Science and Engineering **86** (2015) 012006 doi:10.1088/1757-899X/86/1/012006

Optimization of burning process of hydrocarbon fuels with varying specific heat of combustion

E R Saifullin[1], **Yu V Vankov**[2]

[1]Kazan Federal University, 18 Kremlyovskaya str., Kazan 420008, Russian Federation
[2]Kazan State Power Engineering University, 51 Krasnoselskaya str., Kazan 420066, Russian Federation

E-mail: techphys@kpfu.ru

Abstract. This article explores the combustion of gaseous fuel in the case of an abrupt change of the specific heat of combustion. Analyzes the changes in the rate of heat when the fuel and air flows are constant. Defines the conditions for which retained the initial rate of heat release and optimal fuel-air ratio.

1. Introduction

Optimization of the combustion of gaseous fuels - an urgent task [1], [2], [3]. At present, the associated petroleum gas and waste of refinery plants are often used as a fuel in thermal power plants. These gases are unstable composition and calorific value. During the combustion of such fuels heat release rate will be variable. As a result, the thermal regime of the power plant will be unstable.

In this article it is considered the case of abrupt change of the specific heat of combustion. It is defined as the need to change the fuel and air supply, the heat release rate by the combustion of the fuel remained constant.

2. Initial relations

Use the well-known relations from the theory of combustion processes [4]:

$$q = 7050n + 2500,\ [q] = \text{kcal/m}^3. \quad (1)$$

$$g^* = 7{,}13n + 2{,}28,\ [g^*] = \text{m}^3/\text{m}^3, \quad (2)$$

q – heat of combustion per unit volume of fuel, g^* – the volume of air required for complete combustion of fuel per unit volume, n – carbon number.

Let the fuel has constant over time the heat of combustion q_0. When the complete combustion of the fuel at a rate of $G_{f,0}$ heat release rate is equal:

$$Q_{f,0} = q_0 G_{T,0}, \quad (3)$$

the necessary amount of air flow, according to (2) is equal:

$$G^*_{a,0} = g^*_0 G_{f,0}. \quad (4)$$

Heat flux obtained heat transfer agent related to the rate of heat release by ratio

$$Q_0 = \eta_0 Q_{f,0}, \quad (5)$$

where η_0 – the efficiency of the heat exchanger.

IOP Conf. Series: Materials Science and Engineering **86** (2015) 012006 doi:10.1088/1757-899X/86/1/012006

3. Theoretical analysis

Suppose that at some point in time the heat of combustion abruptly changed by a small amount Δq, and then remains constant over time, i.e.:

$$q = q_0 \pm \Delta q, \qquad \Delta q / q_0 = \theta_q \ll 1. \tag{6}$$

From the relations (1), (2):

$$g^* = k_5 q + k_6, \qquad k_5 = k_3/k_1, \qquad k_6 = k_4 - k_2 k_3/k_1. \tag{7}$$

Given the relations (6), (7):

$$G_a^* = g^* G_{f,0} = G_{a,0}^* \pm k_5 \Delta q G_{f,0}. \tag{8}$$

The (-) sign in (6), (8) corresponds to a reduction of fuel combustion heat. In this case, a complete combustion needed less air than the first.

As a result, the complete combustion of the fuel appears surplus of air, which reduces the combustion temperature [5], [6], [7]. Furthermore, according to formula (3) decreases the rate of heat. Therefore, according to (5) reduced the heat flux obtained by heat transfer agent from the products of combustion.

In order to restore the initial thermal conditions in the complete combustion of fuel, it is necessary to change the fuel and air flows. Let:

$$G_f = G_{f,0} + \Delta G_f = G_{f,0}(1 + \theta_f), \qquad G_a = G_{a,0} + \Delta G_a = G_{a,0}(1 + \theta_a). \tag{9}$$

It is necessary to restore the initial rate of heat, i.e.:

$$qG_f = Q_0 = qG_{f,0}. \tag{10}$$

Terms complete combustion has the form:

$$G_a = g^* G_f. \tag{11}$$

After substitution of (9) into equation (10), (11) and neglecting terms of the 2nd-order terms, obtain:

$$\theta_f = \theta_q, \qquad \theta_a = \theta_f - \theta_g. \tag{12}$$

In view of (6), (7) have:

$$\theta_g = \frac{\Delta g^*}{g_0^*} = -\frac{k_5 \Delta q}{k_5 q_0 + k_6}.$$

Calculations made on the basis of formulas (1) and (2) showed that $k_5 = 10^{-3}$, $k_6 \approx -0{,}25$. I.e. q_0 – quantity of the order 10^4 kcal/m^3 and above, the value $k_5 q_0$ much more than k_6. Therefore:

$$\theta_g = \theta_q. \tag{13}$$

and from (12):

$$\theta_a = 0. \tag{14}$$

The (+) sign in (6), (8) corresponds to an increase of heat of combustion.

In this case, for the complete combustion of fuel unit requires more volume of air.

Because air flow rate remains the same, the increase of the specific heat of combustion of fuel leads to its incomplete combustion.

To restore the initial thermal regime without incomplete burning of fuel changes the fuel and air flows according to the equations (9).

The condition of conservation of heat release rate (10) and the condition of complete combustion of fuel (11) - are the same.

Substituting expressions (6), (8) and (9) into (10), (11), taking into account the linearization of equation (13) gives the following results:

$$\theta_f = -\theta_q, \qquad \theta_a = 0. \tag{15}$$

IOP Conf. Series: Materials Science and Engineering **86** (2015) 012006 doi:10.1088/1757-899X/86/1/012006

4. Conclusions

So, after the reduction of the specific heat of combustion of fuel at a small relative value, to restore the initial thermal conditions can increase fuel flow by the same relative value without changing the air flow.

After raising the specific heat of combustion at a relative small value, reducing the fuel flow to the same relative value without changing the air flow can provide complete combustion and a predetermined value of the heat flow imparted to the heat transfer agent.

The work is performed according to the Russian Government Program of Competitive Growth of Kazan Federal University.

References

[1] Amr I and Saiful B 2007 *Fuel* **87** 1824
[2] Snow D A 2001 *Great Britan B. Plant engineer's reference book* 864
[3] Iovleva O V and Larionov V M 2007 *Russian Aeronautics* **50(3)** 303-308
[4] G. F. Knorre, K. M. Aref'ev, A G Blokh 1966 *Russia B. Theory of combustor processes* 1042
[5] Larionov V M and Nazarenko T I 2000 *Izvestiya Vysshikh Uchebnykh Zavedenij. Aviatsionnaya Tekhnika* **(4)** 69
[6] Larionov V M and Beloded O V 2003 *Izvestiya Vysshikh Uchebnykh Zavedenij. Aviatsionnaya Tekhnika* **(4)** 48-52
[7] Beloded O V and Larionov V M 2006 *Russian Aeronautics* **49(1)** 44-48

Three-Beam Triangulating Sensor

E S Denisov[1], A Sh Salakhova[1], G V Timergalina[1], T P Nikishin[1] and M G Fazlyyyakhmatov[2]
[1] Kazan National Research Technical University named after A.N. Tupolev - KAI, 31 K. Marks str., Kazan 420111, Russian Federation
[2] Kazan Federal University, 18 Kremlyovskaya str., Kazan 420008, Russian Federation

Email: mfazlyjy@kpfu.ru

Abstract. The new high precision triangulating sensor for measuring distance and/or inclination angle with high temperature stability for a wide range of technical and technological applications is proposed. The corresponding measurement algorithm is considered and hardware allowing its implementation is developed. The preferable embodiment of three beam triangulating sensor comprises three laser radiation sources, CCD-array based image sensor including optical system, and control electronic unit.

1. Introduction

Measurement of the distance between objects, or to an object is an actual problem in many industrial and technology applications in geodesy, echolocation, robotics, military applications, etc.

To solve this problem the radio, ultrasonic or optical waves may be used. Measurement of the distance is based on the emission signal in direction of the object and then receiving and processing the reflected or scattered signals.

Ultrasonic waves are widely used in non-destructive testing (NDT) of solid objects, in diagnostics of multiphase objects [1, 2] as well as in of underwater sonar. When measuring the distance or position of remote objects is preferable to use optical or radio waves.

The main advantages of optical sources is a high focusing accuracy, the independence of propagation velocity of the ambient temperature and the relative simplicity of the design.

Optical distance measurement methods can technically be put into three categories: interferometry, time-of-flight and triangulation methods [3].

In time-of-flight pulsed laser rangefinders, a laser pulse is emitted to the target, and range is measured as the time delay between the emitted pulse and the echo received from the target.

This method was used in the following devices [4, 5]. For short-range applications, this technique presents the disadvantage of requiring the measurement of extremely short times for good range accuracy [6]. For example, to obtain 1 mm accuracy, the accuracy of the time interval measurement should be 6.7 ps [7].

The next group is laser phase-shift rangefinders. In a phase-shift rangefinder, the optical power is modulated with a constant frequency. This method is also known as amplitude modulation continuous wave (AMCW). A sine wave of this frequency generated by the main oscillator modulates the dc current of the laser diode. After reflection from the target, a photodiode collects a part of the laser beam. Measurement of the distance is deduced from the phase shift between the photoelectric current and the modulated emitted signal [7]. A couple devises using this principle presented in [8, 9].

Distance measurement by the phase-shift technique provides resolution of some millimeters in 1 to 20 m ranges and with high-frequency modulation the resolution can be better than 50 um. The main disadvantage of this method is the complexity of circuit realization.

Another continuous wave technique uses frequency modulation (FMCW) of the radio frequency subcarrier used to amplitude modulate a laser diode output. In these systems, the range is measured as an intermediate frequency. The main advantage of this technique over AMCW is its ability to resolve multiple targets, which is important in scenes where a main target is obscured by objects in front of it [6, 10].

Some devises provides this method described in [11, 12]. The main disadvantage of this method is that frequency modulation response of a laser diode is nonuniform against the modulation frequency. So that a linear optical frequency sweep cannot be realized by a linear modulation of the control current. In addition, the frequency versus control current characteristic is also in general nonlinear [7].

Many technological and technical application imposes contradistinguished requirements of high accuracy and ease of use, in the field of rangefinders it means that it should have high accuracy and tolerance to misalignment. One of the possible way to meet these opposed requirements is described in the present paper. More specifically, in this paper a method for measuring the distance to the object and its geometrical arrangement in some technical and technological applications using three distributed laser sources is proposed. The additional advantages of this method is possibility to measure inclination angle of the surface under test and to take into account this angle to obtain correct value of distance.

2. Three beam laser triangulating method

In the introduction has been shown that one of the main problem is difficulties caused by requirements of fine alignment of a triangulating sensor in relation to object under test. It is obvious that in many cases such measurements become inconvenient or inaccurate (for example, when object changes orientation during measurement procedures).

To solve this problem three beam laser triangulating method has been proposed (Figure 1). The method uses laser beams generated by three spatially distributed sources (1, 2 and 3). Each beam produces an image on surface of object under test (6) in the form of small dot. The images of three dots are perceived by light sensitive matrix equipped by optical system. Different distances to the object are corresponds to different positions of dots on image similarly to the conventional laser triangulating method [3].

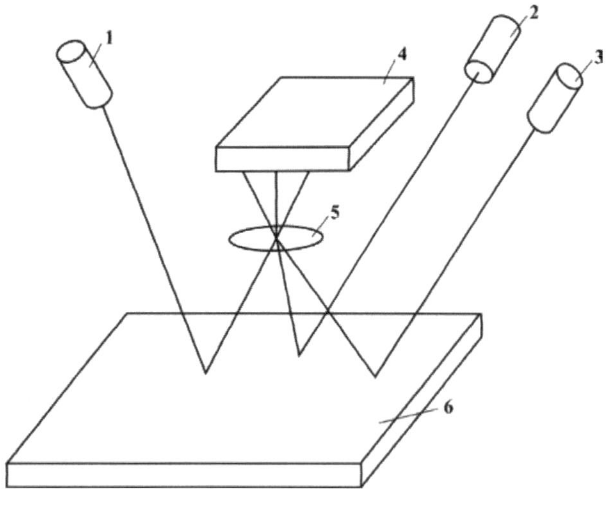

Figure 1. Three beam laser triangulating method: 1, 2, 3 is laser beam sources, 4 is light sensitive matrix, 5 is optical system, 6 is an object under test.

IOP Conf. Series: Materials Science and Engineering **86** (2015) 012007 doi:10.1088/1757-899X/86/1/012007

It makes possible to increase accuracy of distance measurement and to provide measurement of inclination angle of the object when taking into account measurements of distances to an object from three different points with the predetermined positions (laser sources outputs).

Let's consider the proposed method for the two-dimensional case (Figure 2).

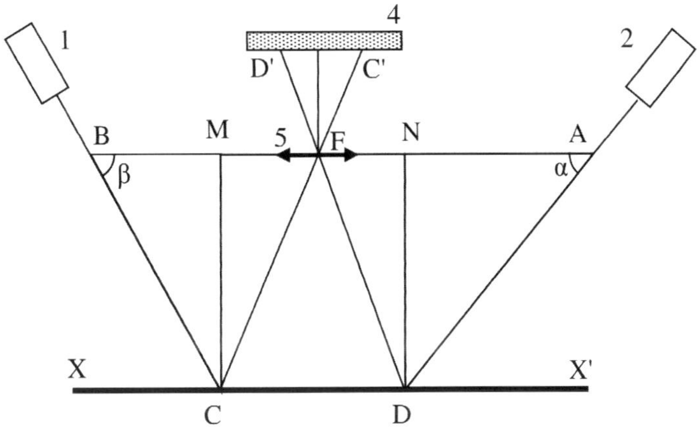

Figure 2. Two-dimensional illustration of three beam laser triangulating method: 1, 2 is laser beam sources, 4 is light sensitive matrix, 5 is optical system.

1) Line AB goes through the external principal plane of the lens system 5 and intersections of the line AB with laser beams are denoted by points A and B.

2) The points of intersection of beams of the laser sources 1 and 2 with plane X-X' of the object under test are designated as C and D correspondingly;

3) Points M and N are projections on the line AB of points C and D correspondingly;

4) F is the optical center of the lens. *f* is the distance from the lens to the image sensor 4 (light sensitive matrix);

5) C 'and D' are images of points C and D on the surface of light sensitive matrix 4;

6) For triangle ADO, we can write

$$\frac{AD}{\sin\gamma A}=\frac{L_A}{\sin(180^0-\alpha-\gamma_A)}, \quad (1)$$

where L_A – length of the segment AB. In this case

$$AD=\frac{L_A\sin\gamma_A}{\sin(\alpha+\gamma_A)}. \quad (2)$$

7) The height ND we find as

$$ND=AD\cos\alpha=\frac{L_A\sin\gamma_A\cos\alpha}{\sin(\alpha+\gamma_A)}. \quad (3)$$

8) Then cathetus ON of a right triangle NDO is equal to

$$ON=\frac{MD}{\tan\gamma_\alpha}=\frac{L_A\sin\gamma_A\cos\alpha}{\sin(\alpha+\gamma_A)\tan\gamma_\alpha}. \quad (4)$$

9) Similarly, for triangle BCO

$$MF = \frac{L_B \sin\gamma_B \cos\beta}{\sin(\beta+\gamma_B)}, \quad (5)$$

$$CM = \frac{L_B \cos\gamma_B \cos\beta}{\sin(\beta+\gamma_B)}. \quad (6)$$

10) The inclination angle of the plane can be found from the following equation (see Figure 3)

$$\tan\theta = \frac{LD}{CL} = \frac{DN - MC}{MN}. \quad (7)$$

These calculations are presented for the two-dimensional case. However, they can be easily generalized to three-dimensional space.

Experimental investigation of the proposed method had been carried out using conventional digital camera Defender G-lens 2693 and three laser pointer has shown that this simple implementation of method allows to obtain instrumental error less than 1% for distance measurement in the range 0,2-2 m, and about 2% for inclination angle measurement at distance 1 m. However, the accuracy can be significantly improved by more precise manufacturing of the sensor and by sensor components selection and adjustment.

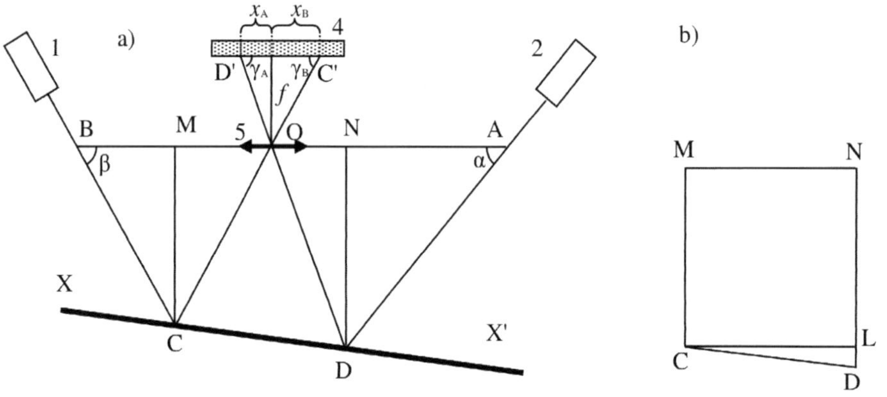

Figure 3. Determination of inclination angle: 1, 2 is laser beam sources, 4 is light sensitive matrix, 5 is optical system.

It is obvious that information on the inclination angle of surface of the object under test could be used for improvement accuracy and usability of sensor. In case of using the proposed method the correct distance to the surface can be measured taking into account the inclination angle of the surface. Further the method provides possibility to not use exact alignment of the sensor since it could make automatic correction procedures eliminating effect of misalignment.

3. Three-beam laser triangulating sensor

The three-beam laser triangulating method described above has effective realization in form of sensor. A proposed three-beam sensor contains three spatially separated laser sources (LS) configured to generate narrow beams, a receiver of optical radiation on the basis of the light sensitive charge-coupled device (CCD) array.

Block diagram of the sensor is presented in Figure 4 according to which sensor comprises three units: Laser Radiation Generator, Light Receiver and Digital Control and Processing Unit. Laser Radiation Generator including three laser drivers (LD) and three laser sources (LS) are adapted to produce three spaced laser beams. Laser Receiver comprises light sensitive CCD-array (LCA), preprocessing unit (PU), control unit (CU), and Memory Unit (MU). LCM forms image with dots corresponding to laser track on the object's surface. PU are intended to perform preliminary processing of image, such as operations of color correction, white-balance, etc. MU stores image before transferring to microcontroller. CU controls operation of whole Laser Receiver. Digital Control and Processing Unit comprises two main units: microcontroller (μC) and interface unit (IU). μC is configured to control operation of whole sensor and to process measuring data, particularly to

- determine positions of beams centers on the image;
- calculate distances to each of laser sources;
- calculate inclination angle of the object under test;
- estimate corrected value of distance between sensor and the object under test.

IU allows to transmit data to computer, global or local network, indicator, etc.

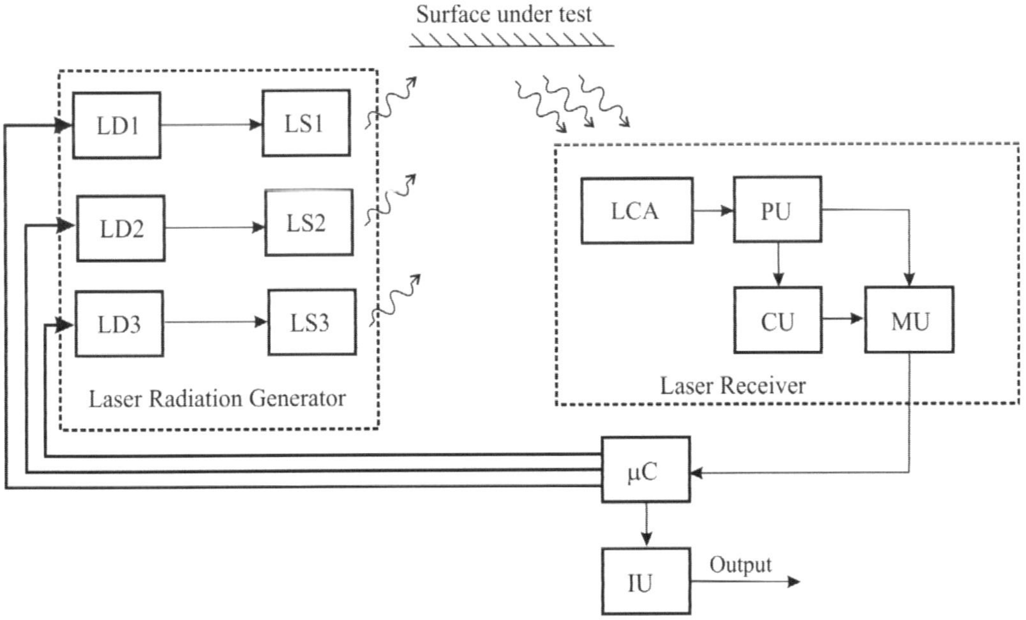

Figure 4. Block diagram of the three-beam laser triangulating sensor.

One of technical problem for this type of sensor is differentiation of beams on the image. Here light sources with different wavelengths (colors) or operation of each LS in its specific period of time (frequency and time separations correspondingly) can be used.

The main advantages of this implementation of sensor is in that it requires not expensive components and the sensor could provide functionality of digital camera between or/and during measurement procedure.

This approach allows to increase accuracy and usability since the sensor is able to take into account inclination angle during distance measuring.

4. Conclusion

The original approach for developing triangulating sensor is proposed. The sensor uses three laser beams to measure distance and inclination angle with excellent temperature stability. This approach allows to increase accuracy and usability of measuring equipment. It can be used effectively in many technological processes (such as control of thickness, distance measurement, etc.), in automotive control systems, in robot orientation systems, and in other applications requiring contactless measurement of distance or inclination angle.

The corresponding measurement algorithm is considered and hardware allowing its implementation is developed. Design of three beam triangulating sensor comprises at least three laser radiation sources, CCD-matrix based image sensor including optical system, and control electronic unit which provides signals for configuring and actuating sensor's elements and processes measuring signals.

To recognize images of different laser beams on the surface of photosensitive CCD-matrix two different methods could be used in dependence on the user and environment requirements: 1. actuating laser-sources in different moments of time; 2. using laser sources with different wave lengths (different colors). Therefore the proposed sensor can be manufactured from inexpensive parts and is very cost effective.

The work is performed according to the Russian Government Program of Competitive Growth of Kazan Federal University.

References

[1] Sagdiev R K, Denisov E S, Evdokimov Yu K, Fazlyyyakhmatov M G and Kashapov N F 2014 *IOP: Conf. Ser. Material Science and Engineering* **69** (1) 012012
[2] Denisov E S, Temyanov B K, Sagdiev R K and Fazlyyyakhmatov M G 2014 *IOP: Conf. Ser. Material Science and Engineering* **69** (1) 012014
[3] Bosch T and Lescure M 1995 *Selected Papers on Laser Distance Measurement*, SPIE Milestone Series, Vol. MS 115 (SPIE Optical Engineering Press, Bellingham, WA) p. 738
[4] Key C, Brumfield J and Baker K 1993 *US Patent №5241360 G01C 3/08*
[5] Abramov A I, Zborovsky A A and Ivanov B B 1997 *RU Patent №2140622 G01C 3/08*
[6] Garcia E and Lamela H 2001 *Opt. Eng.* **40** (1) p 61-66
[7] Amann M, Bosch T, Lescure M, Myllyla R and Rioux M 2001 *Opt. Eng.* **40** (1) p 10-19
[8] Yakovlev Y O, Grigorievsky V I, Astrelin A V, Sherstuk A V and Petrov P M 2006 *US Patent № 6483595 G01B 11/24*
[9] Gogolla T. Winter A and Seifert H 2006 *US Patent №7023531 G01C 3/08*
[10] Stann B, Ruff W and Sztankay Z 1996 *Opt. Eng.* **35** (11) p 3270–3278
[11] Rohling H and Mende R 2000 *US Patent №6147638 G01S 13/36*
[12] Slotwinski A 2006 *US Patent №7139446 G02B 6/00*

Acoustic energy generation by the burning fuel layer

E V Semenova

Kazan (Volga Region) Federal University, Russia

E-mail: techphyskfu@gmail.com, evgeniya.yallina@gmail.com

Abstract. This article studies the process of acoustic energy generation by layer of burning solid fuel in an oscillating air flow. Studied the results obtained for combustion in an open tube. It has been shown that the introduction of the correlation function can be used to determine the acoustic energy generated in the combustion chamber of device capacity-tube type. Received the empirical expression for this correlation function.

1. Introduction

It is known that fluctuations in gas positive effect on the combustion process and the heat- mass transfer in power plants [1-3]. For efficient combustion of solid fuels are widely used devices such Rijke tube. The main problem of theoretical research pulsating combustion in such systems - calculation of acoustic energy generated by the burning fuel layer. Expression defining this energy for burning in an open tube is known. The purpose of this work - to get a similar expression for the combustion of solid fuels in the device capacity-tube type.

2. Relations for open tube

Self-excitation of the gas oscillations in combustion engines due to the generation of acoustic energy in the combustion zone [4, 5], which is

$$A_c = \frac{(B-1)S_1U_{1,0}}{Q_0}\left\langle \mathrm{Re}(p_1^{'})\,\mathrm{Re}(q^{'})\right\rangle_t \quad (1)$$

where B the ratio of the gas temperature behind combustion zone to the temperature of the gas before combustion zone, S and $U_{1,0}$ – the cross-sectional area and mean velocity of the flow entering the combustion zone, respectively, $Q_{1,0}$ and $q^{'}$ - are average and pulsating heat release rate, respectively, $p^{'}$ - the acoustic pressure.

Pulsating rate of heat release associated with the velocity fluctuations of gas before combustion zone by relation $q^{'} = K_u u_1^{'}$ [6, 7]. The transfer function of the flame on the device Rijke tube type (Figure 1a) known [8, 9] and has the form

$$K_u = \frac{\exp(-i\omega\tau_u)}{\sqrt{1+(\omega\tau_1)^2}}, \quad (2)$$

where ω – the angular frequency, τ_1 - the time of inertia, τ_u - the time lag of heat release rate fluctuations with respect to the air velocity fluctuations.

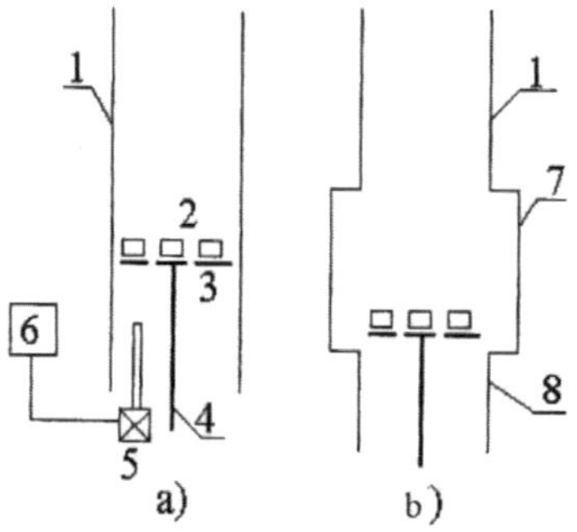

Figure 1. The experimental setup.
a – Rijke tube type, b – capacity-tube type;
1 – resonance tube; 2 – fuel samples; 3 – screen, 4 – stock, 5- microphone, 6 - measuring unit,
7 – combustion chamber - capacity,
8 – air supply tube

3. Correlation function for device capacity-tube type

Another device is shown in Figure 1b. In [8], the diameter of the combustion chamber was more than 3 times the tube diameter. Gas oscillations were excited when the fuel layer located near the outlet air supply tube. There is a minimum and maximum distance between the bottom of the combustion chamber and the fuel layer. In the middle of this interval is the maximum amplitude of the oscillations of gas. The reason is that when the fuel layer is almost over output air supply tube, air velocity fluctuations are small because of the large acoustic impedance. Moving up the fuel layer reduces the resistance and increases the amplitude of pulsations of air velocity to the maximum value. Further increase the distance between the bottom of the combustion chamber and the fuel layer leads to that the airflow is fully expanded in the combustion chamber and pulsating air velocity is reduced to a value of $u_*' = (S / S_c)u_1'$, where S_c - the cross sectional area of the combustion chamber. Amplitude pulsations of air velocity and heat release rate decreases and gas oscillations disappear.

We introduce the correlation function relating the pulsation air to fuel layer with pulsations of air velocity at the outlet air supply tube $u_*' = \psi u_1'$. The expression for the velocity fluctuations of heat has the form $q' = \psi K_u u_1'$, and the formula (1) retains its form.

In [8] the function $\psi = 1 - x_* / x_m$, x_* - a distance from the bottom layer up to the combustion chamber, x_m - a distance corresponding to the oscillation damping. Calculations have shown that the results obtained for small values x_* do not agree with the experimental data. We represent the correlation function in the following form

$$\psi = (S / S_0)(a + b\xi_* + c\xi_*^2 + d\xi_*^3),$$

where $\xi_* = x_* / R_0$, R_0 - the radius of the air supply tube.

According to experimental data [8], for combustion of 6 fuel particles, R_0 =0.02m, S / S_c = 0.1, oscillations are excited when $0.1 \le \xi_* \le 0.6$, and the maximum amplitude of the oscillations corresponds to $\xi_* = 0.3$. Analysis of the total volume of the data showed that $\xi_* = 0.075$ - limit for the excitation of oscillations, that is ψ=0. Other values of the correlation functions were calculated from the relations that determine the conditions of excitation and amplitude of the oscillations [8]: ψ=0.05 for $\xi_* = 0.1$ and 0.6, ψ=0.72 for $\xi_* = 0.3$. This data corresponds to the following correlation function,

$$\psi = S / S_c \left(-6 + 80\xi_* - 186\xi_*^2 + 116\xi_*^3\right), \tag{3}$$

which allows to calculate the acoustic energy generated by burning wood particles layer in the device capacity- tube type with any geometrical parameters.

4. Conclusion.

Because correlation function depends on the geometry and location of layer in the combustion chamber, the expression (3) can be used in other cases, the pulsation combustion of solid fuels. If the transfer function of the flame (2) is determined, the acoustic energy generated by the burning fuel layer in devices such as capacity-tube is according to the formula (1), where the right side is multiplied by the correlation function (3).

The work is performed according to the Russian Government Program of Competitive Growth of Kazan Federal University.

References

[1] Larionov V M and Nazarenko T I 2000 *Izvestiya Vysshikh Uchebnykh Zavedenij. Aviatsionnaya Tekhnika* **(4)** 69
[2] Ermakov R A, Sadykov A F, Galiullin R G, Larionov V M 2011 *Chemical and Petroleum Engineering* **46(11-12)** 749-751
[3] Kashapov N F, Saifutdinov A I and Fadeev S A 2014 *J. Phys.: Conf. Series* **567** 012004
[4] Larionov V M and Beloded O V 2003 *Izvestiya Vysshikh Uchebnykh Zavedenij. Aviatsionnaya Tekhnika* **(4)** 48-52
[5] Beloded O V and Larionov V M 2006 *Russian Aeronautics* **49(1)** 44-48
[6] Iovleva O V and Larionov V M 2007 *Russian Aeronautics* **50(3)** 303-308
[7] Yallina E V , Larionov V M, Iovleva O V 2013 *Journal of Physics: Conference Series* **479(1)** 012017
[8] Larionov V M, Filipov S E, Rukavishnikov D V 2008 *Izvestija Vysshih Uchebnyh Zavedenij. Problems of Energy.* **11-12** 3-15
[9] Semenova E V, Larionov V M, Kazakova E I 2014 *Journal of Physics: Conference Series* **567(1)** 012033

Plasmachemical processing of raw liquid hydrocarbons by sunk microarc-discharge

G,R.Ganieva, B.A.Timerkaev
Kazan National Research Technical University named by A.N.Tupolev, Kazan, 420111 Russia Kazan, K.Marx str., 10

E-mail: btimerkaev@gmail.com

Abstract. The characteristics of microarc discharge with electrodes sunk in the liquid hydrocarbons were investigated. The gas bubble formation process in the space between the electrodes was analyzed. The method of decomposition of heavy hydrocarbons to lighter fractions in plasma of micro-discharge with sunk electrodes is proposed.

In this paper we describethe method of deeper processing of raw liquid hydrocarbon materials for the production of light fractions and allotropic forms of carbon. As shown in [1 - 4] the microarc discharge between electrodes sunk into material is idealfor this purpose. In this case, the discharge is ignited directly into gases that are released from the raw liquid hydrocarbons and occupy a space between the electrodes. Microarc discharge with sunk electrodes was selected for hydrocarbons processing due to several factors. Firstly, discharge takes only limited area inside the liquid. Secondly, small microdischarge current and small energy input don't heatthe gas up to the high temperatures when the raw hydrocarbonscouldsplitinto atomic state, and then atoms could combine in certain fractions. Thirdly, there is no need for electrodes coolingas they are sunk in liquid. Also instanthardening for getting target products are not required, as this process occurs naturally at the gas-liquid boundary.

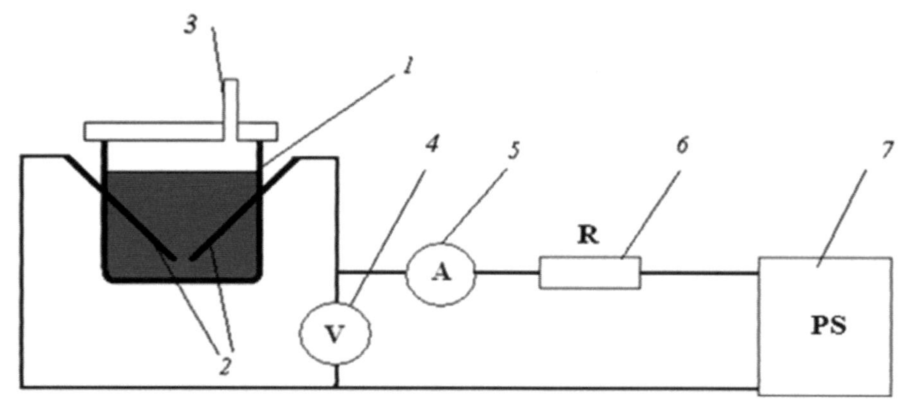

Figure 1. Schematic diagram of the experimental setup: 1 - container with the raw hydrocarbons, , 2 - electrodes, 3 - gas exhaust tube, 4 - static voltmeter, 5 - ammeter, 6 –ballast resistance box, 7 - the power supply.

The laboratory setup for the decomposition of the raw hydrocarbon by microarc discharge is a ceramic crucible 1 with lid tightly closed. The lid has holes for the input and output of electrodes 2 and gas exhaust tube 3. The high-voltage regulated power supply 7 allows to submit the rectified voltage up to 15 kV. Theballast resistance 6 is connected to electrical circuit. The fuel oilwas used as rawhydrocarbon material. Two copper electrodes 2 of 4 mm diameter are immersed in fuel oil.

To initiate of electric arc it was necessary for short time to connect the electrodes, and then separate them by a distance of a few millimeters (1-2 mm). High temperature of electrical microarc in the space between the electrodes providesenough pressure to supportplasma area ofthe gas-vapor bubbleinside the fuel oil. The boundary of this area is directly contacted with fuel oil. Due to the high temperature the high-boiling fractions of heavy hydrocarbons are in the plasma area and by impacts of fast electrons and high energy ions they are splitted into smaller fractions. Due to the action of buoyancy force gas-vapor bubble expands and floats.The decomposition of heavy hydrocarbons and the formation of light fractions in the chamber (between the surface of the fuel oil and lid) lead to pressure growth and the gas starts to leak out through a special exhaust tube into the prepared container.

During its movement in the liquid the bubbleis reduced in volume.This could be caused by two phenomena. First, boiling and vaporization if the investigated liquid soccured only in a limited areabecauseof theirlow thermal conductivity, so the vapours of hydrocarbon raw materials was absorbedback by theliquid. Second, the floating gas-vapour bubble was been cooled and its volume was reducing.

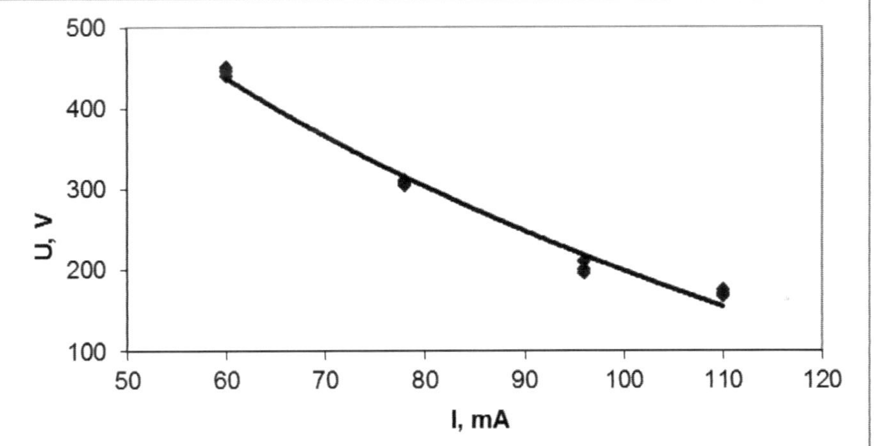

Figure 2. Current-voltage characteristic of the electric microdischarge in fuel oil.

The high viscosity and low thermal conductivity of the fluid enables the existing of extensive conductive region which expands since thedischargepower is growing. The shape of these characteristics suggests that the conductive gas-vapour region increases in proportion to the discharge current. Falling volt-ampere characteristic of microdischarge in fuel oil (Fig. 2) is evident that the discharge is a high voltage arc discharge. At low currents the voltage reaches 0.5 kV. If we consider that the inter electrode distance is of the order of 1 mm, the reduced electric field E/N in that discharge is of the same order as in the glow discharge. Therefore, the mechanism of formation of charged particles is not purely thermal and is partly shock ionization. The existence ofsuch mechanism of formation of charged particles and the presence of high-energy electrons in the plasma allow the use of high voltage arc for the decomposition of long hydrocarbon molecules in the fuel oil into shortermolecules.

The temperature of the arc depends on the current values per unit cross-sectional area of the electrode i.e. current density. The larger it is, the higher the temperature of the arc is. Herewith heat sink from microarcis performed by free convection. However, at such high temperatures the molecules dissociate, that require more energy input. The pressure in area of electrical microarc sunk in fuel oil

IOP Conf. Series: Materials Science and Engineering **86** (2015) 012009 doi:10.1088/1757-899X/86/1/012009

corresponds to the depth of immersion of microarc. Since the depth of the fuel oil is just a few centimeters so the pressure is about atmospheric value.

Under such conditions, unlike electric microarcs in fuel oil from the free microarcs will be the following. Arc channel will be filled with gases and vapors of hydrocarbons of diverse factions. That part of the fuel oil that is in direct contact with microarc will be boiling, put in the discharge region various petroleum fractions. Surface boiling and relatively low thermal conductivity of the fuel oil are preventing the overheating of the bulk of fuel oil and its coking. Hydrocarbon molecules, being in the area of electrical discharge are attacked by fast electrons and ions of the discharge, as well as the excited atoms and molecules of hydrocarbon gases. As a result, most of the long hydrocarbons will be decomposed into shorter fractions. Formed hydrocarbon gases and vapors quickly leave the discharge region because of the effects of buoyancy forces and heat, giving place to the new flow of hydrocarbons. Hydrocarbons vapours consisting a great quantity of gasoline and other light fractions are dissolved in fuel oil, gradually turning it back into the oil. The gaseous fraction is also partially absorbed in the fuel oil, but most of them break out. After collecting of this gas it has a white color and it is heavier than air, despite the fact that its contains methane and hydrogen.

Microarcs electric power, which is released in the discharge,spends to heat the fuel oil,to heat the gas,to vaporization, to decomposition of heavy hydrocarbons, as well as to heat the chamber. Also, the processing of fuel oil by the sunk microarc forms coke deposits on the electrodesbesides getting gas. Chromatographic analysis of released gases showed that half of them is ethylene, 36% is hydrogen and about 7% is methane.

References

[1] Timerkaev B A, Ganieva G R, Ziganchin D I and AuhadeevM M May-June 2014 Electrical microdischarges in liquids and prospects of their application in plasma chemistry *Journal of Engineering Physics and Thermophysics* Vol 87, №3 (Minsk) pp 677-681

[2] Arslanov A Sh, Zalyaletdinov F D, Zakirov IM and Timerkaeva DB 2010 Decomposition of hydrocarbons in the arc plasmaflow *Vestn. KSTU Tupolev* №3 (Kazan) pp 137-142

[3] Ganieva G R, Galeev I G, Gismatullin N K, Ziganshin D I, Tahautdinov R Sh and Timerkaev B A 2011 The decomposition of heavy hydrocarbons in the free electric arc *Izv. Samara scientific.Center RAS* Vol 13, №4 S pp 1156-1159

[4] Ganieva G R, Galeev I G, Gismatullin N K, Ziganshin D I and Timerkaev B A 2012 The decomposition of heavy hydrocarbons in a recessed arc. *Vestn. KSTU Tupolev* № 4 (Kazan) pp 184-188

Systematization of accuracy indices variance when modelling the forming external cylindrical turning process

I P Balabanov, L A Simonova, O N Balabanova

Kazan Federal University, Naberezhnye Chelny Institute
423810, Naberezhnye Chelny, Mira av., 68/19

E-mail: balabanovip@mail.ru

Abstract. The article considers the problem of accuracy deviation systematization for external cylindrical turning, proposed a hierarchical approach to the evaluation of these deviations, an approach to the analysis of nesting accuracy metrics, as well as, the common scheme of identification of deviations of the accuracy metrics for party billets in external machining were proposed.
Key words. modeling of the forming process, accuracy metrics hierarchy, identification of deviations, batch work pieces.

1. Introduction

The precision prediction can be attributed to the most important tasks of technological design in today's environment. Objective forecast is required for process parameters selection for unused equipment, and analysis activities to improve the quality of required parameters [1].

The designer specifies the forming accuracy requirements for various parts of even moderate complexity by several tens of indicators. Accuracy metrics of details are regulated by GOST [2]. The analysis of the content has shown that the definitions are focused only on conformity assessment. The entire set of indicators cannot be represented in a single system. None of them expressed in a vector, so if you measure each indicator separately, it is impossible to evaluate them geometrically. However, each of the indicators, one way or another, affects the quality of performance part assignment function. Therefore, for example, a statistical analysis of the process accuracy by one single indicator may not be effective. The more complete description of the set of indicators, the more effective management will be.

Introduction of the quality management system at the enterprises involves the use of processes monitoring, which is necessary to establish quantitative measurable indicators [3] [4]. These include indicators of product quality and process parameters. Therefore, it was decided to use the maximum amount of to obtain a more complete characterization of the process to the task of modeling.

2. Basic part

The goal of modeling is as follows: identifying opportunities for improving the accuracy of the products and reduce the cost of run production jobs processes on the jobs of morphogenesis. The result of the simulation is representation of vectors in each measure of accuracy of all details that are generated during execution of the job in the workplace in a single coordinate system; the distribution of their values in the party, the dynamics of the cost of the process taking into account all additional operations [5].

IOP Conf. Series: Materials Science and Engineering **86** (2015) 012010 doi:10.1088/1757-899X/86/1/012010

For solving the problem the assumption was formulated that, in accordance with the contents of the details forming process, the accuracy of each element of the workpiece is characterized by the integral (total) index, which consists of hierarchically nested into each other constituents (nesting is represented in Figure. 1).

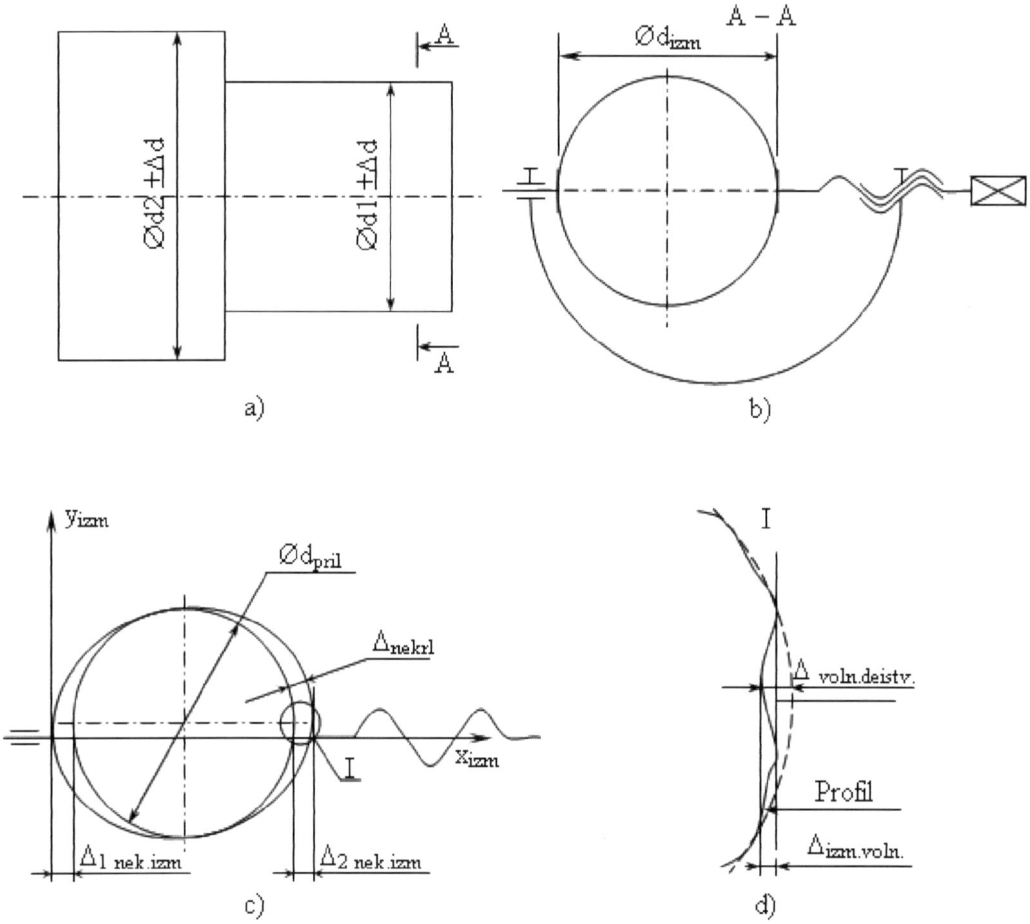

Figure 1. Scheme of components deviation of the measured diameter of the cylindrical element:
a) size designation in the drawing;
b) scheme of the two-point measurement;
c) element of the deviation profile;
d) element of the deviation waviness.

As follows from the figure, the appointing of engineer a single measurement to the diametric size point-to-point method will have the following results:

- the actual value of the diameter (Fig. 1. b.) is actually estimated, as the profile deviation and waviness are ignored due to errors in processing, which the designer assumes within the tolerance;
- measured values are derived from the diameter of the adjacent circumference (Fig. 1. c.) deviations of the profile from the adjacent and deviation of the waviness profile (Fig. 1. d).

If the angular coordinates of the profile deviations and the waviness are not specified, the contribution of these deviations in the measurement results is unknown. To sum them up geometrically, it is necessary to introduce a common system of coordinates.

In order to simulate the formation process deviations, it is necessary, moreover, to represent the measured indicator of accuracy of the detail manufactured item as a consequence of successive overlapping components of the total impact factors during the production cycle.

So, scallop's roughness is a consequence of single tooth contact of the tool with the amount of the allowance. Its value depends on the parameters of the movements, geometry – on the geometry of the tooth in the plan. Core complex billet excessive roughness can cause wrinkling scallops, although symmetrical clamping of elements will not affect the position of the workpiece [6].

Roughness is a part of waviness. Waviness is a periodic deviation from the adjacent line profile. Technologically it is a consequence of the radial or end play of the teeth of milling cutters and grinding wheels, imbalance circles, beating spindles. It must be taken into account when the workpiece positioning the prisms, the cartridges with a small number of clamping elements and a small area of the contact points. Waviness is one of the components of the deviation profile either in the transverse or longitudinal direction.

The profile deviation (nonroundness, crookedness) is a consequence of fluctuations in the allowance due to the deviations of the workpiece, error-based, insufficient rigidity of the movable joints, and non-rigidity of the tool. Each of them contributes to the deviation of the shape.

Size deviation in cross section of detail manufactured item, measured in 2-point scheme, in each point is the sum of the deviations that occur during morphogenesis [7]:

- module adjustment deviation;
- settings for the size deviation;
- adjacent deviation from configuration due to insufficient rigidity;
- shape deviation due to the non-rigidity of the technical system;
- waviness deviation due to the runout of the grinding wheel.

The next levels of the hierarchy are crookedness of the element items axes and the deviation of the relative position of the axes.

The nested variance analysis requires a single technological system of coordinates that characterize the machine tool system as a whole. Usually it is considered to be self-sufficient. But to trace the link between the conditions and the actual value of the deviations, both in morphology and in the measurement, it should be used in both of these processes.

Based on the objectives, inputs include the following initial information:

in the workplace: code of the workplace; the scheme of machine tool system composition; data on the accuracy of its individual modules; the calculation of the initial position of the modules in the processing coordinate system; selecting controls.

Actual production tasks are performed in a planned restrictions (set optimization process will be done with their account):

- Planned quantity of good parts.
- Acceptance criteria.
- Performance incentives for compliance with quality assurance.
- Required time of delivery of the products.
- Products delivery schedule.
- Incentives indicator for adherence to the delivery schedule.
- Cost limit of job processing
- Incentives indicator for cost reduction.
- The number of blanks in the party.

Next, data for the preparation and enforcement of basic operation production jobs is entered:

- getting machine tooling; input accuracy parameters and time consumption;
- receiving tool; input accuracy parameters and time consumption;
- the receipt of funds control; input time consumption;

- base snap-in; input time consumption;
- calculation of the position of the tooling in the machine coordinate system;
- base tool; input time consumption;
- setting the size in the coordinate system of the machine; input time consumption;
- processing and measurement test parts; input time consumption;
- planning cycle control and regulation control operations;
- coordinate system calculation of the tool in the machine coordinate system;
- calculation of time consumption to complete the operation.

Within a single task accuracy indices of the elements of the machining system are usually considered systematic. Each blank, at the same time, based randomly, and because they are numerous and they are often not identified, it is considered that the indicators of the accuracy of the workpieces are simply unknown. Actually, even if these values are not known to the technologist, in fact, each workpiece is inherent in the full set of indicators. When measuring, they just become known, and identified. Therefore, for efficient simulation, at least random values of each individual parameter that affects the final accuracy of the machining of the part must be specified. These include all accuracy metrics, as well as the hardness and strength of the material machined parts. A condition is made in the task that each of them can meet the requirements of the operating map, but in real life doesn't have to match. Depending on the processing technology in the previous operations, each of them is characterized by a well-defined distribution law in the party with the largest scalar value. And also a well-defined distribution of deviations in the coordinates of the workpiece is typical.

Identification of the party involves procurement structuring and development of indicators for each component. The structure of the workpiece when processing each set consists of supporting and processing complexes. The complexes include separate elements that define a single coordinate plane, and the elements will include structural components, each of which defines the coordinate of a single point of contact.

The first stage of identification is allocation to individual blanks of the highest values of the scalar variance of the indicators, starting with the simplest ones (waviness, profile, size of the adjoining, the axes deviation). Distribution is carried out in the following order:

The source data formation:

- job matching tolerance;
- the share of non-compliant blanks;
- the highest value of the deviation;
- the expectation of the variance in party relative to the middle of the field of tolerance;
- the value of Sigma;
- choice-of-law distribution;
- building distribution curve.

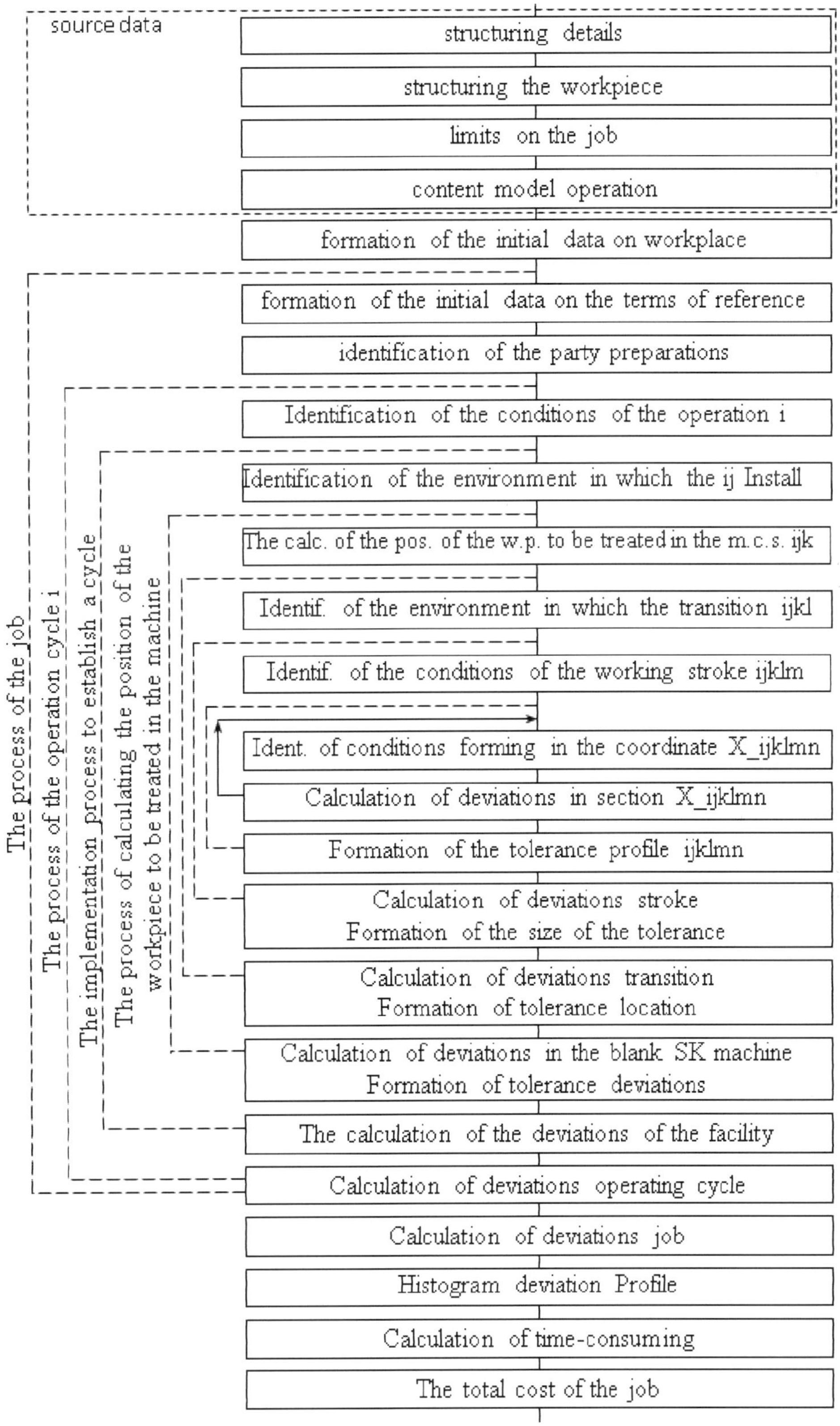

Figure 2. Scheme of modeling process of forming the party preparations in the workplace.

Histogram plotting [8] [9] [10]: intervals formation taking into account the division value of the measuring means; determination of the spacing of the party blanks; calculation of the frequency of exposure interval;

The second stage is the identification of the deviation vector in the coordinate system of the workpiece [5]: setting the coordinate system of the workpiece (origin, axis direction); technological constraints on variations of the distribution of deviations input; setting coordinates of the largest vector value (specific or random distribution); distribution equation construction of variations in workpiece coordinate (if necessary).

According to the results of the identification information model of a single workpiece is formed. Prepared system of the source data is used in the modeling process (Figure 2).

3. Conclusion

Each structural element of the operation model allows to calculate the amount and position deviation. At the end of all forming operation simulation, distribution of the largest (smallest) vectors values of all the variances in the volume of the batch according to the normal Gaussian distribution is formed.

The scope of the technique:

- the most effective method for learning tasks (learning system errors, processes optimization, knowledge testing);
- application techniques in plant cash-flow problems:
- direct tasks (accuracy prediction when entering the dimensions in the technological system; the opportunity to influence the process of manufacturing parts and provide the required parameters taking into account the individual characteristics of the machine);
- inverse tasks (parts, equipment and tooling state diagnosis; analysis of the minimum quality requirements of the party preparations to ensure the required quality details).

References

[1] Pan. Y.Z., Ai, X., Tang, Z.T.: Optimization of tool geometry and cutting parameters based on a predictive model of cutting force (in Chinese). China Mechanical Engineering 19, 428-431 (2008)

[2] GOST 24642-81 Basic norms of interchangeability . Tolerances of form and position . Basic terms and definitions . official publication

[3] M. Standartinform 2012 . GOST ISO 9000-2011. Interstate standard . Quality management system. Fundamentals and vocabulary (ISO 9000:2005, IDT) Quality management systems. Fundamentals and vocabulary (enacted on January 1, 2013 Order Rosstandart from 22.12.2011 N 1575 - st)

[4] M. Standartinform 2012 . GOST ISO 9001-2011. Interstate standard . Quality management system. Requirements (ISO 9001:2008, IDT) Quality management systems. Requirements (came into effect from January 1, 2013 Order Rosstandart from 22.12.2011 N 1575 - st)

[5] SV Kasyanov , LA Simonov. IP Balabanov Managing the process of forming the party details using a simulation model / / Problems of mechanical engineering and materials technology at the turn of the century: Collection of articles VIII International Scientific and Technical Conference. Part I. - Penza, 2003, p. 198-201

[6] Bajic. D., Lela. B., Zivkovic. D.: Modeling of machined surface roughness and optimiza-tion of cutting parameters in face milling. Metalurgija 47.331-334 (2008)

[7] Balabanov IP, Simon LA Modeling precision molding processes based on the identification of indicators of the quality of a blank / / Proceedings of the III International scientific-practical conference " CAR AND Technosphere " , Kazan , 17-20 June , the Kazan : Kazan State Publisher University of Technology, 2003 , 1160 p

[8] Magnus YR, Katishev PK, Peresetskii AA, Econometrics . Initial course . Textbook. 2nd ed . , Rev. - Moscow: Delo , 1998 . - 248 .

[9] Cover Thomas M. Elements of Information Theory. — John Wiley and Sons, 2006. — P. 254.
[10] (2009) «Maximum Entropy Autoregressive Conditional Heteroskedasticity Model». Journal of Econometrics (Elsevier): 219–230.

Formation of initial data of the workpiece batch in simulation modelling precision forming

I P Balabanov, O N Balabanova, A V Groshev

Kazan Federal University, Naberezhnye Chelny Institute
423810, Naberezhnye Chelny, Mira av., 68/19, Russia.

E-mail: balabanovip@mail.ru

Abstract. Modern simulation model is limited to the creation of items. However, in actual production, we often have significantly more manufacturing products. In the article the question of the formation of the initial batch data is raised. This will more accurately simulate the processing of the party details.

1. Introduction

Using ISO 9000 series sets, new tasks appeared for introducing them in the manufacturing enterprise. [1; 2]. Today it is not enough to ensure compliance with the specified values of all indicators of the volume of output, but also to stabilize the processes required to improve them [3; 4]. The basis for such controlling is able to be the simulation.

In engineering, the following measures of accuracy are given [5]: the relative position, the total deviation of the shape and the relative position, the shape deviations, waviness, roughness. After the analysis of definitions and measurement schemes it was discovered that their application is directed strictly to determine the deviation limits; there is no requirement to commit coordinate measurement; figures do not constitute a single coherent system; the part is not considered as a set of interrelated elements. Nevertheless, each of the indicators somehow affect the quality of the assignment functions. Therefore, the complete description of a set of indicators is necessary for the better management model [6].

2. Basic part

In the frame of single task indicators accuracy of the tool elements is usually considered to be systematic. Each part, at the same time, is random, and because of their quantity they can't often be identified, so it is considered that measures of accuracy workpieces are unknown [7]. But in fact, even if these values are of unknown technology, in reality full set of indicators is inherent in every workpiece. When measuring they just become known. Therefore, for the effective simulation at least random values of each individual indicator must be set which may affect the overall accuracy of the workpiece. These include all measures of accuracy, as well as the hardness and strength of the material of the workpiece. In the task the condition is that each of them can meet the requirements of the operating map, but in a real situation they may not. Depending on the processing technology on previous operations each of them is characterized by a well-defined distribution law for the largest party in the scalar value. A well-defined distribution law of deviations in workpiece shows that coordinates are typical. In this paper, we face the task of obtaining the most complete information over the status of each workpiece simulation accuracy for forming the party details on the basis of standard accuracy metrics. Let's look at an example of party identification workpieces made in accordance with the requirements of the drawing (see Figure 1). Material: calibrated, incised, cold-rolled round steel (GOST 7417-85).

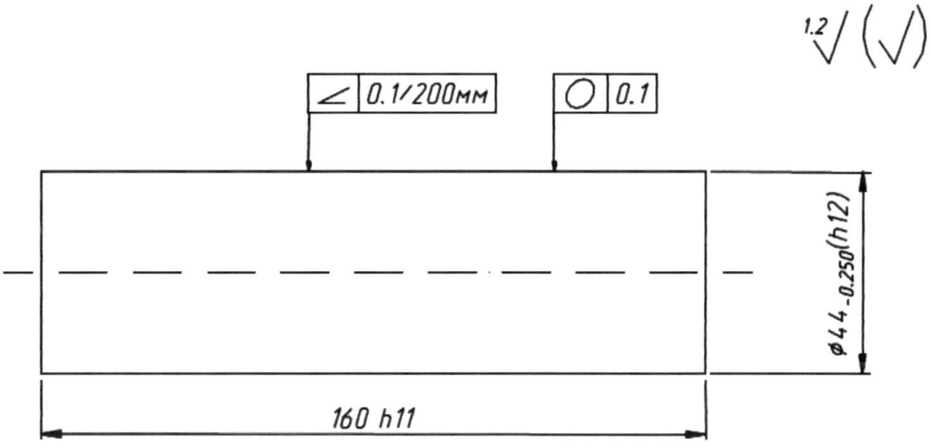

Figure 1. Model limitations of standard accuracy metrics

The initial data for the party preparations, according to the drawing, will serve: material grade; the hardness of the workpiece (the law of distribution over the surface of a workpiece); diametrical rod size (d44-0.250); deviation from circularity (0.1); conicity workpiece (0.1/2000 mm); workpiece (160-0.250).

For the full simulation process of morphogenesis full identification of the status of all elements of the system is required. This element is also a workpiece. In accordance with the content of the process of shaping parts of each element of the workpiece, accuracy is characterized by the integral (total) index, which consists of hierarchically nested components.

It is also necessary further to define the conditions of installation and handling. In this example, it includes the definition of the end of the workpiece fastened in a holder and operator skills. In the first case, the position of the workpiece is identified in the global coordinate system of the machine, in the second - qualification will allow the operator to simulate the precise adjustment of the cutting tool size.

Each set in the drawing accuracy rate workpieces party is characterized by its distribution law in the party. For example, the distribution of the diametric dimension (d44-0.250), deviations from circularity (0.1), conicity of the workpieces(0.1/2000 mm) length of the workpiece (160-0.250), hardness and skill of the operator on (the limb) are subject to the normal distribution (Gause) and fixing in the chuck - equal probability. The material trademark can be considered constant.

At the first stage of identification party must assign each billet unique number constant throughout the workpiece life cycle. Next, we form a system of initial batch data for each of the workpieces of proposed indicators

Allocation algorithm in more detail is shown by the example of figure diametrical size of the workpiece. As mentioned above, the distribution of the diametrical size of the workpiece in the party subject to the normal distribution law. The initial data for the construction of such a law would be: number of pieces in the game (N); nominal value of distributed parameter (dnom); upper deviation of the accuracy (ES); lower deviation of the accuracy of (ei); width of the interval (sigma); centre grouping (X); [8; 9]

In the second stage, it is necessary to translate the histogram distribution curve. To do this, ask the sampling interval, i.e. determine the width of the column histogram [8; 10]. It is advisable to take the sampling step equal dividing the price of the measuring tool (for measuring 0.025 micrometer).

$$f(m) = \frac{1}{\sqrt{2\pi} \times \delta} \times \exp\left(-\frac{(x-m)^2}{2\delta^2} \right) \tag{1}$$

where δ - the coefficient of the sample standard deviation; x - the value of shifting the center of the grouping.

Then determine the values of the distribution function at the midpoints of sampling intervals by formula (1) and calculate the sum of the values of the distribution function (2) at the midpoints of sampling intervals:

$$B = \sum_{m=ei}^{es} f(m), \tag{2}$$

where ES - upper deviation of the accuracy; ei - the lower deviation of the accuracy.

To complete the calculation of the amount of detail in the discretization interval it is necessary to calculate the distribution coefficient index, i.e. number of units of a normal distribution attributable one part in the batch (3):

$$k = \frac{B}{N} \tag{3}$$

where N - number of pieces in the batch;

and then determine the number of pieces relating to each interval, by dividing the distribution function in the middle of each sampling interval on the distribution coefficient of batch (4).

$$n_i = \frac{f(m)_i}{k} \tag{4}$$

In connection with estimated quantities workpieces on the intervals to the whole, there can be a situation where the total value of workpieces is different from the number of workpieces in the party. To resolve this discrepancy it is proposed to compare the amount of workpieces obtained in histogram to the original value, and add / remove missing / excessive harvesting of the center column grouping, thereby bringing the number of workpieces in accordance with the initial data.

At the final stage randomly assign each room a specific workpiece value of the index exactly, in accordance with this distribution.

3. Conclusion

In accordance with an algorithm, accuracy is defined by specific values of all other parameters. After the completion of the identification of the values of all parameters, we obtain the accuracy of available state database of each workpiece. The values obtained are fully consistent with regulations for batch of workpieces. This information will create a model of each particular workpiece based on law describing each part in the machine system, a move equation of the workpiece relative to the machining part of a cutter and as a result, to obtain the equation removes allowance.

Thus, thanks to this technique, random errors and processing bases, due to inaccuracy of manufacturing can be determined regularly.

References

[1] M. Standartinform 2012 . GOST ISO 9000-2011. Interstate standard . Quality management system. Fundamentals and vocabulary (ISO 9000:2005, IDT) Quality management systems. Fundamentals and vocabulary (enacted on January 1, 2013 Order Rosstandart from 22.12.2011 N 1575 - st)

[2] Chow-Chua, Clare; Goh, Mark; Wan, Tan Boon (2003). "Does ISO 9000 certification improve business performance?". International Journal of Quality & Reliability Management 20 (8) : 936. doi: 10.1108/02656710310493643.
[3] M. Standartinform 2012 . GOST ISO 9001-2011. Interstate standard . Quality management system. Requirements (ISO 9001:2008, IDT) Quality management systems. Requirements (came into effect from January 1, 2013 Order Rosstandart from 22.12.2011 N 1575 - st)
[4] Corbett, Charles J.; Montes-Sancho, María J.; Kirsch, David A.; Song, CX; Wu, H; Dai, Q; Irier, H; Upadhyay, AK et al. (2005). "The Financial Impact of ISO 9000 Certification in the United States: An Empirical Analysis". Management Science 51 (7): 1607-16 . doi: 10.1287/mnsc.1040.0358. JSTOR 20110397 . PMC 3292193 . PMID 22037496 .
[5] GOST 24642-81 Basic norms of interchangeability . Tolerances of form and position . Basic terms and definitions . official publication
[6] SV Kasyanov , LA Simonov. IP Balabanov Managing the process of forming the party details using a simulation model / / Problems of mechanical engineering and materials technology at the turn of the century: Collection of articles VIII International Scientific and Technical Conference. Part I. - Penza, 2003, p. 198-201
[7] Balabanov IP, Simon LA Modeling precision molding processes based on the identification of indicators of the quality of a blank / / Proceedings of the III International scientific-practical conference " CAR AND Technosphere " , Kazan , 17-20 June , the Kazan : Kazan State Publisher University of Technology, 2003 , 1160 p
[8] Magnus YR, Katishev PK, Peresetskii AA, Econometrics . Initial course . Textbook. 2nd ed . , Rev. - Moscow: Delo , 1998 . - 248 .
[9] Nancy R. Tague (2004). "Seven Basic Quality Tools". The Quality Toolbox. Milwaukee, Wisconsin: American Society for Quality. p. 15. Retrieved 2010-02-05.
[10] Wheeler, Donald J. "What About Charts for Count Data?". Quality Digest. Retrieved 2010-03-23.

Cross-flow and heat transfer of porous permeable cylinder

I V Morenko[1], V L Fedyaev[1,2] and E R Galimov[2]

[1] Institute of Mechanics and Engineering, Kazan Science Center, Russian Academy of Sciences, Russia
[2] Kazan National Research Technical University named after A N Tupolev, Russia

E-mail: morenko@imm.knc.ru

Abstract. Non-isothermal flow around the porous permeable cylinder of square cross section by a viscous incompressible fluid is considered. The motion and energy equations is solved numerically using the finite volume method. The character of the fluid flow, change of the cylinder drag coefficient and Nusselt number as a function of Reynolds and Darcy numbers is analyzed.

1. Introduction

Porous permeable bodies streamlined by fluid are widely present in nature and engineering particularly in biological filters, cooling systems, heat exchangers, nuclear reactors and dryers.

In virtue of the practical relevance, complexity of hydrodynamic and thermal processes the problems of flowing around the porous permeable bodies are of great interest. First studies of fluid flows through a porous medium were carried out by N.Ye. Zhukovsky [1]. In the 50-80s of the last century flowing around the bodies was studied analytically [2]. Recently, with the development of methods of mathematical modeling and computer technologies the numerical methods are increasingly used [3-7].

In this paper we investigate the non-isothermal flow around the porous permeable cylinder of square cross section by a viscous incompressible fluid at moderate Reynolds numbers Re=1÷40.

2. Governing equations

As a computational domain we choose rectangle with length 0.5 m and width 0.2 m wherein at a distance of 0.1 m from the inlet section at an equal distance from lateral boundaries streamlined cylinder with cross section in the shape of a square with the side a =0.01 m is located.

In the Cartesian coordinate x_1Ox_2 system the beginning of which is placed in the center of the cylinder section axis Ox_1 is directed along the incident flow, plane-parallel laminar flow of a viscous incompressible fluid is described by the equations

$$\frac{\partial v_i}{\partial x_i}=0, \tag{1}$$

$$\frac{1}{\varepsilon}\frac{\partial v_i}{\partial t}+v_j\frac{1}{\varepsilon^2}\frac{\partial v_i}{\partial x_j}=-\frac{1}{\rho}\frac{\partial p}{\partial x_i}+\nu\frac{1}{\varepsilon}\frac{\partial^2 v_i}{\partial x_j \partial x_j}-\frac{\nu}{k}v_i,\ i=1,2. \tag{2}$$

Here t is the time; $v_i = \varepsilon u_i$ is the velocity components of fluid filtration, u_i is the velocity vector components $\vec{u}$, p is the pressure, ρ is the fluid density, ν is the coefficient of kinematic fluid viscosity, k is the cylinder permeability, the porosity

$$\varepsilon = \begin{cases} 1 & \text{outside the body,} \\ 0 < \varepsilon < 1 & \text{within the body;} \end{cases}$$

summation is performed over the repeated indices.

Energy conservation equation for a fluid has the form

$$\frac{\partial}{\partial t}(\rho E) + \nabla\left(\vec{u}(\rho E + p)\right) = \lambda \Delta T - Q_w, \qquad (3)$$

where $E = h - \frac{p}{\rho} + \frac{u^2}{2}$, h is the enthalpy, T is the fluid temperature, λ is the thermal conductivity coefficient of a fluid, Q_w is the capacity of the heat flow to the frame of a permeable cylinder.

Velocity profile, pressure, temperature of the fluid are specified in the inlet section of the computational domain. "Soft" boundary conditions meaning the alignment of both hydrodynamic and thermal characteristics of the fluid flow at a distance from the streamlined cylinder are set at the outlet of the computational domain. On the lateral boundaries – symmetry conditions. At the initial time $t = 0$ fluid instantaneously starts to move.

The problem is solved by the finite volume method.

3. Results

We consider the motion of a fluid with the following parameters (air): density ρ=1.225 kg/m³, dynamic viscosity coefficient $\mu = 1.79 \cdot 10^{-5}$ kg/(m·s), specific heat capacity at constant pressure $c = 1000$ J/(kg·K), thermal conductivity coefficient λ=0.0242 W/(m·K). The air temperature on inlet $T = T_0$=300 K, the temperature of the cylinder frame T_w=350 K.

When fluid flows around the body the main calculated values are drag coefficient $C_D = \frac{F_{x1}}{0.5 \rho u_\infty^2 a}$, Nusselt number $\mathrm{Nu} = \alpha a / \lambda$, which characterizes the intensity of heat transfer. Here u_∞ is the fluid longitudinal velocity at the inlet of the computational domain, F_{x1} is the projection on axis Ox_1 of the force of hydrodynamic resistance, $\alpha = \frac{\lambda}{T_w - T_\infty}\left(\frac{\partial T}{\partial n}\right)$ is the local coefficient of convective heat exchange, n is the outward normal to the cylinder surface. Average over the cylinder surface Nusselt number $\overline{\mathrm{Nu}}$ is also calculated.

During numerical experiments we found that the flowing around the porous cylinder of square cross section at said Reynolds numbers and porosity $\varepsilon = 0.4 \div 0.9$ occurs in the stationary mode, vortex trail does not form. For small numbers of Darcy Da=k/a^2 flow around pattern of the permeable cylinder is similar to the flow around pattern of impermeable cylinder (figure 1 *a*). With the increase of values of Darcy number the fluid passes through the body. In this case when Da=10^{-4}, 10^{-3} the vortex region behind the body is preserved (figure 1 *b*, *c*), when Da=10^{-2} it disappears (figure 1 *d*).

The drag coefficient of the cylinder decreases with an increase of both Reynolds and Darcy numbers. For calculating its value in the range $5 \le \mathrm{Re} \le 40$, $10^{-6} \le \mathrm{Da} \le 10^{-2}$ it is proposed to use the approximation formula

$$C_D \cong 16.0(1 - 0.07 \lg \mathrm{Da}) \cdot (1 + 0.06 \mathrm{Re}) / \mathrm{Re}.$$

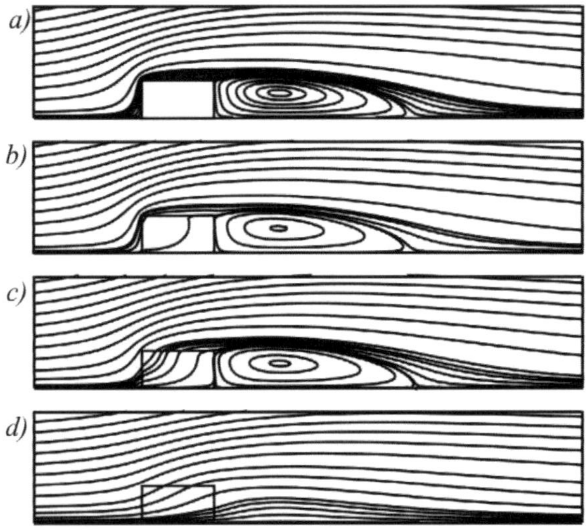

Figure 1. Streamlines on the half of the computational domain when Re = 40 and Darcy numbers Da: *a*) 10^{-6}, *b*) 10^{-4}, *c*) 10^{-3}, *d*) 10^{-2}

Figure 2 shows the variation of Nusselt number at different parts of the cylinder surface. It can be seen that heat transfer is mainly carried out on the windward side of the cylinder on the rear side it is minimal. With the growth of cylinder permeability Nusselt number on its windward side of noticeably increases, but on the lateral and rear sides, conversely, slightly decreases.

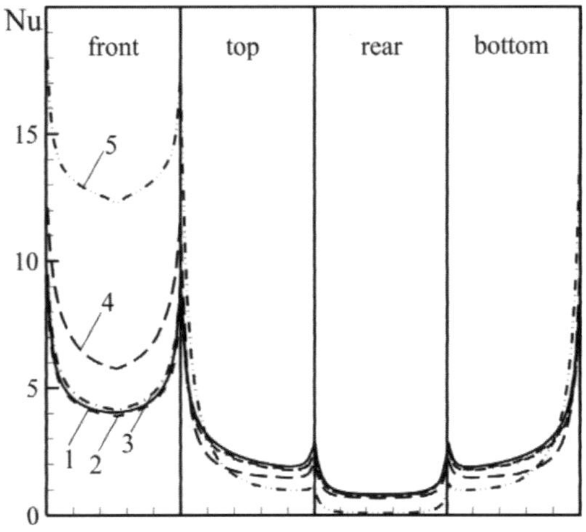

Figure 2. Local Nusselt number when Re=40 on the cylinder surface (front, top, rear, bottom sides one after another): 1 – impermeable, 2 – permeable, Da=10^{-6}; 3 – Da=10^{-4}; 4 – Da=10^{-3}; 5 – Da=10^{-2}

Comparing the obtained values of the average Nusselt number $\overline{\mathrm{Nu}}_f$ on front side of the streamlined cylinder with the data [6, 7], we see that for small permeability (Da=10^{-6}, Da=10^{-4}) of the cylinder, they are practically the same (figure 3). In case of greater permeability (Da=10^{-2}) when Re>20 the discrepancy is observed and it increases with the growth of Reynolds number. In our opinion, this discrepancy is due to the fact that in work [6] only the surface of the frame of streamlined body is heated, whereas in this work - the whole frame. Furthermore, takes into account the inertial component depending on the pressure drop when fluid flows through the permeable body which can be better seen at higher Reynolds numbers [6].

In contrast to the front side on the rear side of the cylinder surface heat transfer decreases with an increase of permeability (figure 4). It is interesting that, when the Reynolds numbers increase and the Darcy numbers are small (Da=10^{-6}, Da=10^{-4}, Da=10^{-3}) the average Nusselt number $\overline{\mathrm{Nu}}_r$ on this side of the cylinder surface increases, notably the increase is the most intensive at Re <5. In the range $5 \leq \mathrm{Re} \leq 40$ dependence of $\overline{\mathrm{Nu}}_r$ on Re is almost linear.

In case when Darcy number is relatively high Da=10^{-2} and Re is approximately equal to 5 there is an obviously expressed maximum of dependency of $\overline{Nu}_r$ on Re. Reduction of Nusselt number with the decrease of Reynolds number is quite understandable, whereas deterioration of heat transfer with an increase of Reynolds number when Re>5 should be explained. As can be seen from figure 1 *d* the vortex region is not formed behind the cylinder, the fluid flow rate through the cylinder is relatively high in the central part and low at the periphery. Significant unevenness of fluid velocities in cross section of the cylinder transversal to the flow, in our opinion, can be the reason for this unusual behavior of the Nusselt number $\overline{Nu}_r$ as a function of the Reynolds number of the incident flow.

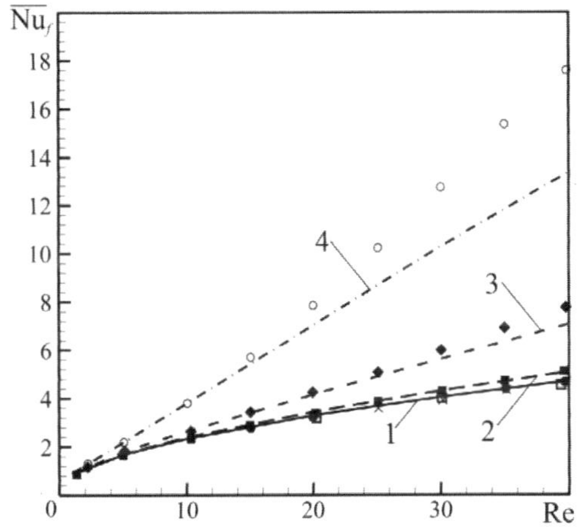

Figure 3. Average Nusselt number $\overline{Nu}_f$ on the front (windward) side of the cylinder surface when Re=40 and values of Da: 1 – 10^{-6}; 2 – 10^{-4}; 3 – 10^{-3}; 4 – 10^{-2}. Comparison with the data of other authors: ● – 10^{-6}, ■ – 10^{-4}, ♦ – 10^{-3}, ○ – 10^{-2}, × – impermeable cylinder [6]; □ – 10^{-6} [7]

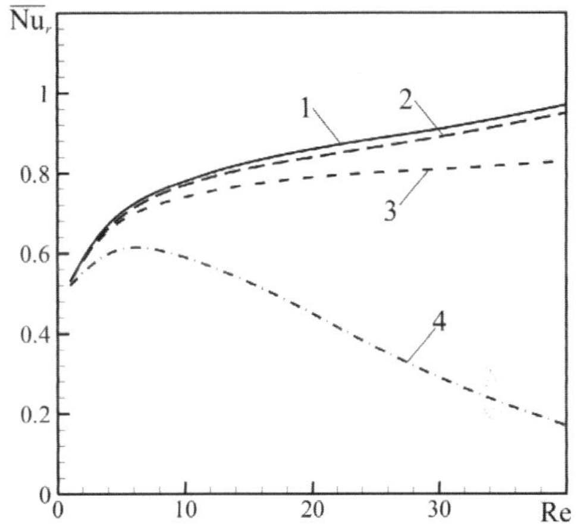

Figure 4. Average Nusselt number $\overline{Nu}_r$ on the rare side of the cylinder surface when Re=40 and values of Da: 1 – 10^{-6}; 2 – 10^{-4}; 3 – 10^{-3}; 4 – 10^{-2}

As it appears from the results of calculations on the lateral surfaces of the cylinder average Nusselt number $\overline{Nu}_b$ changes depending on the Reynolds and Darcy numbers as follows

$$\overline{Nu}_b = 0.45(1-0.022\lg Da)\cdot(1-0.008Re)\sqrt{Re}.$$

Average Nusselt number for the entire cylinder surface

$$\overline{Nu} = 1.25+0.15(1+0.125\lg Da)\left(Re^{-5}\right).$$

When carrying out numerical calculations of flow around the porous permeable cylinder of square section by a viscous incompressible fluid for moderate Reynolds numbers it was determined that with increasing Darcy number a pair of vortex regions behind the body disappears, the drag reduces, the heat transfer increases, mainly on the front surface. As follows from the analysis of the calculated data the approximation dependencies for the drag coefficient of the cylinder, average Nusselt number for the entire surface and Nusselt number for the lateral surface were obtained. It is shown that usage of porous permeable bodies in thermotechnical plants intensifies the heat transfer due to the development of the contact surface; this reduces the hydrodynamic resistance as compared to solid bodies.

References

[1] Zhukovsky N Ye 1937 *Water seepage through the dam* (Moscow: Complete edition) 363

[2] Stechkina I B 1979 Resistance of porous cylinders in a viscous fluid flow for low Reynolds numbers *Izv. AS USSR. MZhG* **6** 122–124
[3] Yu P, Zeng Y, Lee T S, Chen X B and Low H T 2011 Steady flow around and through a permeable circular cylinder *Computers & Fluids* **42** 1–12
[4] Bhattacharyya S, Dhinakaran S and Khalili A 2006 Fluid motion around and through a porous cylinder *Chemical Engineering Science* **61** 4451–4461
[5] Jue T 2004 Numerical analysis of vortex shedding behind a porous square cylinder *Int J Numer Methods Heat Fluid Flow* **14**(5) 649–63
[6] Dhinakaran S and Ponmozhi J 2011 Heat transfer from a permeable square cylinder to a flowing fluid *Energy Conversion and Management* **52** 2170–2182
[7] Sharma A and Eswaran V 2004 Heat and fluid flow across a square cylinder in the two-dimensional laminar flow regime *Numer. Heat Transfer Part A* **45** 247–269

Processing methods with imposing of electric field at low-waste division of materials

I.I. Khafizov

Kazan (Volga Region) Federal University, Russia

E-mail: khafizov@kpfu.ru

Abstract. Due to everything extending practice of electrochemical processing of firm alloys the important role is got by studying of their electrochemical behavior. Management of the combined process of probably independent change of parameters of a chemical and mechanical component with restrictions of their limiting values.

1. Introduction

Procuring operations on division of all types of materials include manual and machine cutting on the equipment of different function. Are for this purpose used as traditional ways (the press, the metal-cutting equipment with the metal and abrasive tool, etc.) and new types of processing (the laser is sharp, electroerosive division, ultrasonic processes). In mechanical engineering there is a reliable information about the mastered methods, their limiting possibilities and shortcomings. With increase in a share of expenses at materials there was a problem of research of new types of division of materials, especially it concerned scarce and expensive alloys of type of precious metals, tungsten, magnetic alloys, fragile semiconductors where the exit of suitable details after processing became less than a half of initial weight, and the defects brought in a blanket at cutoff, remained in a product and reduced its characteristics

In world practice division of materials generally carry out the reinforced disks, an electroerosion, an ultrasonic method, the laser. However, such methods don't allow to provide demanded rather high requirements for accuracy, quality of a blanket, productivity, decrease in losses of a material. Besides, the majority of applied methods causes environmental pollution (a dust when using an abrasive, etc.)

Combination of various impacts on object of processing allows to design the combined methods, in particular electroabrasive (electrodiamond). These methods are applied for are sharp when receiving preparations with the subsequent processing which in some cases (manufacturing of details of devices, radio engineering, control facilities) is undesirable since leads to unjustified losses of a material, to secondary errors and defects, sharply increases cost of products. Establishment of unequivocal communications between properties of processed materials, a combination of impacts of the combined process allows to create the modern automated equipment with management of a mechanical, chemical, erosive component in the uniform process, providing receiving after division of materials of ready details 30 microns with a margin error aren't higher and with a roughness 0,32 microns aren't higher. Negative impacts on environment are thus eliminated and to 2 times the cycle of manufacturing of details is accelerated.

IOP Conf. Series: Materials Science and Engineering **86** (2015) 012013 doi:10.1088/1757-899X/86/1/012013

Use of similar processes accelerates creation of new competitive products, expands technological possibilities of production, promotes decrease in deficiency and costs of materials. It is actual for modern mechanical engineering and meets world requirements to new production.

The combined methods of processing are directed on an intensification of process of anode dissolution. Speed will eat metal and accuracy of a formoobrazovaniye at electrochemical processing depend on that, there will be how fast a reaction of transition of a material of preparation in шлам. Speed of anode dissolution is limited to existence of the film passivating a surface, and thickness of a diffusive layer which is overcome by deleted products of processing.

At electroabrasive grinding (abrasive grains or a filler) eliminate firm particles a film, activating thereby process of electrochemical processing. The sizes of abrasive grains defining an interelectrode gap, as a rule, don't exceed the tenth shares of millimeter. At such small gaps the density of a current will be much more, than in case of dimensional electrochemical processing. Speed metal sharply increases in an area of coverage of abrasive grains of the tool. Besides, the part of an allowance is removed mechanical grinding. Unlike usual grinding at anode and abrasive processing on a surface of preparation stronger riveted layer isn't formed, and productivity of grinding raises. Therefore, intensity removal metal at anode dissolution increases owing to mechanical removal of a passivating film and acceleration of process of carrying out of products of processing from an interval, and electrochemical dissolution of a part of metal, in turn, promotes increase of speed of mechanical grinding. Except the specified components removal at small gaps electroerosive process can take place. At the small sizes of a gap the part of metal of preparation is removed at the expense of an electric erosion [2].

For more complete use of advantages of electrochemical processing it is necessary to project details taking into account features of process of anode dissolution of alloys. It is necessary to consider that at electrochemical processing there is no division into draft and fair operations - at any mode to electrochemical processing the height of roughnesses corresponds to fair operations of machining, and with increase of speed removal metal the roughness of a surface decreases. Unlike machining technological indicators to electrochemical processing even raise with increase in hardness of a material of preparation. Besides, at electrochemical processing the tool either doesn't wear out at all, or wears out inappreciable (at the combined way of processing).

Dimensional electrochemical processing considerably expands technological possibilities of manufacturing of details. Thanks to it it is possible to receive forms of the surfaces which creation by other ways or is impossible, or it is unprofitable.

Applied methods of division of metals allow, generally to carry out procuring operations where high precision and quality of a blanket which are provided at the subsequent stages of processing demanding considerable allowances on process, having high labor input and specific power consumption isn't required.

The analysis of known processes and the equipment shows that it is possible to reach high precision of details at division at the expense of establishment of regularities of process under variable conditions of processing, creation of automated control systems by process with adaptation of parameters, in particular tool giving – a disk, managements of its condition at are sharp, control and correction of provision of a cutting part in a groove.

The second effective direction of researches on decrease in losses of scarce materials is equipment use with deduction of details before the completion of calibration of lateral surfaces of a groove.

Prospects of use of results of work in other branches applying precious and scarce materials (a medical technology, stomatology, control facilities devices, electric sockets, etc.) where the economy from elimination of losses of metals can make the considerable sum are opened.

For more complete use of advantages of electrochemical processing it is necessary to project details taking into account features of process of anode dissolution of alloys. It is necessary to consider that at electrochemical processing there is no division into draft and fair operations - at any mode to electrochemical processing the height of roughnesses corresponds to fair operations of machining, and with increase of speed removal metal the roughness of a surface decreases. Unlike machining technological indicators to electrochemical processing even raise with increase in hardness of a material of preparation. Besides, at electrochemical processing the tool either doesn't wear out at all, or wears out slightly (at the combined way of processing).

2. Conclusions

Thus, increase of accuracy of processing at the minimum losses of a material at division is possible when using regularities of process of division by disk tools with settlement geometrical parameters at adaptive management on several coordinates, creation of new ways of management by the provision of various parts of the tool and groove calibration after preparation division with the minimum allowance.

The exception of the subsequent operations on processing of places of division of details allows to accelerate several times processing process, considerably to increase accuracy and quality of details, to lower an expense of scarce materials.

References

[1] Khafizov I I, Sadykov Z B 2007 Management of process of the combined (electrodiamond) processing of highly rigid materials *Vesntik Kazan state technical university named after A N Tupolev* 1 18-20

[2] Khafizov I I, Sadykov Z B 2012 The analysis of ways of low-waste division of scarce materials with imposing of electric field *Vesntik Kazan state technical university named after A N Tupolev* 2 97-100

[3] I.A.Konahina, I.R.Gil'manshin, T.R.Safin. Heightening of efficiency fuel using installations due toexternal recuperation units of waste heat // *IOP Conference Series: Materials Science and Engineering*, Bristol-UK. Vol.69, Conf.1, December 2014. (SJR(2013): 0.130; IPP(2013): 0.139; SNIP(2013): 0.152; DOI: 10.1088/1757-899X/69/1/012009)

[4] I.R.Gil'manshin, I.A.Konahina, N.F.Kashapov, N.N.Fahreev. Mini-Central heating and Power Plant (CHP): the choice of the optimal structure and modes of operation // *IOP Conference Series: Materials Science and Engineering*, Bristol-UK. Vol.69, Conf.1, December 2014. (SJR(2013): 0.130; IPP(2013): 0.139; SNIP(2013): 0.152; DOI: 10.1088/1757-899X/69/1/012008)

[5] I.R.Gil'manshin, N.F.Kashapov. Energy service contracts in regional engineering center for small and medium businesses // *IOP Conference Series: Materials Science and Engineering*, Bristol-UK. Vol.69, Conf.1, December 2014. (SJR(2013): 0.130; IPP(2013): 0.139; SNIP(2013): 0.152; DOI: 10.1088/1757-899X/69/1/012010)

[6] N.F.Kashapov, I.R.Gil'manshin, I.A.Konahina. System analysis of the energy complex of engineering enterprise as a basic tool of effective energy management // *IOP Conference Series: Materials Science and Engineering*, Bristol-UK. Vol.69, Conf.1, December 2014. (SJR(2013): 0.130; IPP(2013): 0.139; SNIP(2013): 0.152; DOI: 10.1088/1757-899X/69/1/012024)

[7] I.A.Konahina, N.F.Kashapov, I.R.Gil'manshin, R.R.Ganiev. Optimization of industrial steam supply and steam-and-condensate farming of machine building enterprise// *IOP Conference Series: Materials Science and Engineering*, Bristol-UK. Vol.69, Conf.1, December 2014. (SJR(2013): 0.130; IPP(2013): 0.139; SNIP(2013): 0.152; DOI: 10.1088/1757-899X/69/1/012025)

Innovative technologies of waste recycling with production of high performance products

I.R. Gilmanshin[1], A.V. Ferenets[2], Yu.I. Azimov[1], A.I. Galeeva[3], S.I. Gilmanshina[1]

[1] Kazan (Volga region) Federal University, Russia
[2] Kazan National Research Technical University named after A.N. Tupolev - KAI, Russia
[3] Kazan State Power Engineering University, Russia

E-mail: is-er@yandex.ru

Abstract. The innovative ways of recycling wastes as a tool for sustainable development are presented in the article. The technology of the production of a composite material based on the rubber fiber composite waste tire industry is presented. The results of experimental use of the products in the real conditions. The comparative characteristics of the composite material rubber fiber composite are given. The production technology of construction and repairing materials on the basis of foamed glass is presented.

1. Introduction

The concept of effective and sustainable development is including the development of raw materials with high performance characteristics and the formation of a closed life cycle [7]. Additional incentive to research in the field of composite materials and waste treatment technologies is systematic increase of primary raw materials cost [8, 9]. The waste management in Russia for a long time has not been given much attention, in the result the problems of landfills overload and waste sorting are occurred, when in meanwhile the major amount of wastecould be used as the raw materials for the production of high-quality innovative products.

The target of using new types of raw materials is being given to modern production, such as composite materials, which could be ensure the production of products with high technical performance, physical and mechanical properties [4, 10]. Rubber mechanical products must have high wear resistance, durability, ensure the long life of mechanisms [4, 5]. In particular, obtaining a composite of fiber-filled rubber compound ensures the production of rubber products, such as wedge transmission belts with high heat resistance and durability of the resource.

2.Governing equations

Wedge belts which are used in the power and energy mechanisms and units are given requirements such as heat resistance and high resistance on wearout. The requirements for heat resistance and wear resistance are defined by the mechanisms work conditions, in which wedge belts are exposed to high mechanical stresses with dynamically changing parameters, with a high friction effect, bending moments of stretching and compression, causing internal stresses in the material with heat and rapid wearout effects on the surface of the material friction [1, 6]. Nowadays abroad to increase heat resistance for rubber (100-120 ° C), chloroprene caoutchoucs are used with the addition of special types of natural caoutchouc, which causes a significant cost of product. To ensure the strength parameters for wearoutsin the technology of rubber mechanical products in recent years the use of composite rubber compounds filled with short fibers are being given greater interest [2, 3]. Short fibers, which are anisotropic particles, are able to navigate in the elastomeric materials. This makes it possible to directionally regulate the properties of the rubber quality control, in particular, on the wear resistance.

The currently developed innovative technology of extrusion obtaining of fiber-filled composite mixture is giving the possibility to use secondary resources of industrial production - waste tire production (e.g. cutting parts of rubberized cord JSC "Nizhnekamskshina"). At the same time the problem of waste utilization and prevention of solid waste disposal is solving. Taking in account that the volume of this waste is 500-600 tons per year, the economic effect of the technology introduction at JSC "Nizhnekamskshina" could be up to 30 mln. rubles per year.

The experiments' results on the use of the obtaining composite for the manufacture of rubber fiber composite have showen that the raw material guarantees the high technical and operating characteristics of the final products, which is confirmed by field tests of wedge belts type C-4000 on the oilpumpers of oilfield NGDU "Zainskneft" of JSC "Tatneft".

The pilot line of wedge belts, installed with the staffing at oilpumper by 4 pieces after a work time about 5000-5500 hours, later have been transferred to the complete set of 2 pieces, which gave the opportunity to continue testing the strength of the belt in a double load to the total 8500-8700 hours.

Data on the properties of rubber fiber composite (RFC) and rubber compounds of isoprene caoutchoucSKI-3 with the introduction of the RFC is determined by the following indicators:

• The minimum torque on the rotor of the rheometer according to the content in the rubber mixture RFC varies slightly, which substantially defines the viscous consistency of the mixture.

• Curing speed is reduced by 25-30%.

• Cohesive strength at high levels in the RFC in rubber composition is rising upsharply,adhesion is increasing to steel cord.

• Tear resistance increases significantly.

• Hardness of rubber mixture by Defoe is increasing and elongation of vulcanizates is decreasing sharply.

• Relative strength is being reduced by 30-35%.

Experimental data are proved by suggest guidelines for opportunities of the effective use of RFC in rubber elements and parts of machines and mechanisms of transport, agricultural and industrial equipment, which are subjected to high stress abrasion and wear.

In terms of ensuring energy efficiency of the production process is of interest to the innovative technology of production of decorative thermal insulation material of construction on the basis of foam glass with high performance [11, 12].

Glass waste is one of the main components of municipal solid waste (MSW). It could cause negative environmental impact by processes of leaching certain hazardous substances and open-pit dumping poses a serious threat to wildlife. The proportion of glass waste in the process of re-cycle is unacceptably low, however it could be used for different purposes.

Re-cycle of glass waste is reducing the energy consumption by more than 30%. A range of products allowing the inclusion of recycled glass is quite wide.

The proposed technology relates to the production of porous silicate foams as decorative foam glasstiles used as cladding, thermal insulation, high strength construction material.

Products of porous silicate foam (foam glass, ceramic foam, glass foam grit), as well as products made on the basis of light porous aggregates, due to their high rates of heat and sound insulation for simultaneously with sufficient strength characteristics found widespread use in the conduct of thermal barrier coatings of building envelopes having significant advantages relatively shielding with existing technologies in terms of reliability, durability resource, technology.

It is known the method described in the book "Foam glass" by Demidovich and some other way of preparation of raw materials mixture, a glass melting furnace, heating granulation, mixing the granules with blowing agent, grinding the batch, peaking in the tunnel furnace annealing product.

Disadvantages of these methods are defined by complexity and energy inefficiency of technological process which are including the aggregate of complex multi-stage operations with the considerable processing time and low productivity.

The proposed method makes it possible to produce a wide range of construction and finishing materials with high energy-saving performance.

3. Conclusions

As result, waste recycling is the way of human impact reduction on the environment, energy consumption of production processes and the formation of approaches to sustainable development. The products made from the recycled materials not only inferior, but often outperform counterparts made from raw materials.

References

[1] Repeated vulcanizates of rubber crumb. Thematic review TsNIITEneftehim / OG Polyakov, AM Chaykun, Moscow, 1993. P. 30.

[2] Makarov V.M., Drozdovskii V.F. The use of amortization tires and waste production of rubber products. -A. : Chemistry, 1986., Leningrad branch.

[3] Solovyov E.M., Zakharov N.D. Processing and use of waste tire industry. M., TsNIITEneftehim, 1986.P. 66.

[4] Zachiesova G.N., Stallions A.N., Zinoviev L.A. Preparation and use of fine powders and powder rubber reclaim. M. TsNIITEneftehim, 1987. - P.60.

[5] V.N.Yaroshevsky. Recovery and recycling of used tires abroad. - M., TsNIITEneftehim, 1994. - P. 84.

[6] Shahovets S.E., Bogdanov V.V. Integrated regeneration tires. - St. Petersburg: "Prospect of Science", 2008 P.192.

[7] The processing and use of waste tire industry. Thematic review. - M., TsNIITEneftehim, 1983. Solovyov E.M., Zakharov N.D.

[8] Nanotechnology in the production of foamed glass building materials // Construction of a new generation, new technologies - new equipment. - №1.- 2010.- 15-19.

[9] Puzanov S.I., Kets A.A. Complex processing of cullet in the production of building materials // Ecology and Industry of Russia. - №12.- 2009.- 4-7.

[10] A.I.Puzanov, S.I.Puzanov, A.A. Ketov. Utilization of Glass Cullet for the Production of Binding Materials // Role for Concrete in Global Development: Proceeding of the International Conference / Dundee. United Kingdom, 2008, P. 317-324.

[11] TECHNOlith insulating concrete - The ecologically green shade of gray. Betonwerk und Fertigteil-Technik/Concrete Plant and Precast Technology, Volume 78, Issue 4, 2012, Pages 62-63

[12] G.Betti, U.Pinori, A.Marradi. The use of recycled glassfoam aggregates for lightweight embankment // Sustainability, Eco-Efficiency and Conservation in Transportation Infrastructure Asset Management - Proceedings of the 3rd International Conference on Tranportation Infrastructure, ICTI 2014 / Pisa, Italy , 2014, P. 245-254.

Development of methodology for controlling the parameters of TP

K V Klochkova, S V Petrovich, L A Simonova, L R Yusupov
Kazan Federal University, Naberezhnye Chelny Institute
423810, Naberezhnye Chelny, Mira av., 68, Russia.

E-mail: kvsimonova@mail.ru

Abstract. This article describes TP control step of the intelligent system for predicting the properties of CGI, which includes three parts: the selection of parameters for comparison, the comparison with the simulation results, the change of the current TP. The list of parameters under which control in the production is carried out has been determined, the adjustment algorithm of TP is designed.

1. Introduction

TP control phase is needed in case of a mismatch of intermediate values obtained during the modeling operations TP and control results in the production, in this case TP and chemical composition adjustment is performed by introducing additional elements (elementwise) or modifiers. If adjustment is required only for one element, the rule base is used [1]. If there is a mismatching of several elements, the simulation is carried out by predicting using a cascade neural network. The TP adjustment is necessary because there is currently a large percentage of deviations in technology - alloy overmodification, non-observance of temperature indicators and incorrect choice of modifiers to adjust the composition.

Intelligent system [2] should provide a preparation of the optimal process, depending on the data entered by the operator. To compile TP we should have all the necessary data on the chemical composition of the product, its mass and wall thickness, and the allowable ranges of technical and economic indicators. If the chemical composition or ranges of technical and economic parameters are not known, it is necessary to predict them and only after that start the TP compiling.

2. Basic part

TP control phase is divided into three blocks: selecting parameters for comparison, a comparison with the results of modeling, the change of the current TP. First, the list of parameters by which a production control is carried out has been defined. After determining the parameters pairs for comparison are formed, consisting of TP parameters values in the manufacture and the values of the simulated TP.

Comparison of these pairs allows you to clarify whether the process has been deviated from the simulated TP, and if it is so - by how much. If you deviate from the planned process it will change. Changing of the TP is carried out by means of the rule base, the genetic algorithm and the neural networks [3]. With the help of the rule base the possible actions by which the TP parameters in the production will match the simulated TP are identified . Genetic algorithm based on abovementioned rules, allows you to choose the best way to solve this problem, taking into account the possible combinations of TP parameters.

Compilation of TP involves selection of the base cast iron or its spectral and thermal analysis to obtain the necessary data to select and calculate the number of modifiers. Rules for selecting base cast iron must have the following logical chain: "Basic cast iron products for producing the products with compacted graphite cast iron should have: a coefficient of carbon equivalent in the range of 4-4.7, 3.5-3.9% carbon content, oxygen content - 500-100 ppm and etc.".

Next, using the rule base, processing method, a casting method and the choice of heat treatment are selected. Such rules should have the following logical chain: "When modifying the base cast iron modifier FSMg6PZM1,5 with magnesium content of 6.3%, 1.5% REM and calcium - to 0.4% was used. Since this modifier refers to the comprising a high amount of magnesium in its composition there should be applied an intra-form ladle inoculation. The maximum length of the article is 700 mm, the minimum wall thickness of 5 mm, weight 42kg - so chill casting is the most appropriate way of casting. "

It may be several possible TP , the most optimal solution is selected among them, depending on the quality, cost, availability of equipment and raw materials, etc.

To ensure the process it is carried out: preparation of charge and ladles, melting and refinement of the initial cast iron melt according to the required chemical composition, equalizing of the initial cast iron melt, the calculation of the required amount of active nodularizing , vermiculating (magnesium, REM), alloying (tin, copper, molybdenum, nickel) and graphitizing (silicon, aluminum, alkaline earth metals) elements based on the content of the initial melt cast iron the core elements (carbon, silicon, etc.), oxygen and sulfur, weighing of ligatures, modifier, metal processing, chemical analysis, the temperature and the time count , transportation of cast iron, deslagging, sampling on the physico-mechanical properties, casting [4].

Analysis of the sources [5], [6], [7] showed that quality control of raw materials, intermediate products of various degrees of readiness and finished product includes (Figure 1):

- Incoming inspection of materials for compliance with regulatory documentation;
- Express control of the chemical composition of the melt on a certain set of elements;
- Control of the melt temperature;
- Control of the propensity of iron to whiten;
- Visual inspection of casting.

Also in the sources [8], [9] is mentioned that in the development of TP of manufacturing castings sometimes additionally are carried out:

- Thermal analysis of the melt behavior during cooling and crystallization;
- Manufacturing and analysis of technological samples for the determination of the casting properties of the melt;
- Production and analysis of samples for determining the mechanical and special consumer properties of the alloy;
- Non-destructive testing of bulk and surface defects in castings.

The production process is based on monitoring of the whole TP operation by operation, according to the modeled TP. The function of the intellectual system is as follows: if a mismatch of intermediate values obtained in the simulation operations of TP and testing results in the production, in this case an adjustment of TP and chemical composition by introducing additional elements or modifiers is carried out.

Adjustment of the chemical composition is made:

- Using a rule base;
- Using a cascaded neural network [10];
- Using predictive models.

TP adjustment is performed according to the algorithm shown in Figure 2.

Changing of the current process is carried out by means of rules, a cascade of neural network and a genetic algorithm. With the help of the rule base volumes of TP parameters application of which will allow to achieve the compliance with modeled TP are defined. Simulation is performed after correcting using cascaded neural network. Genetic algorithm can find the optimal solution, taking into

Figure 1. Controlled parameters in determining the quality of the CGI products on the TP stages

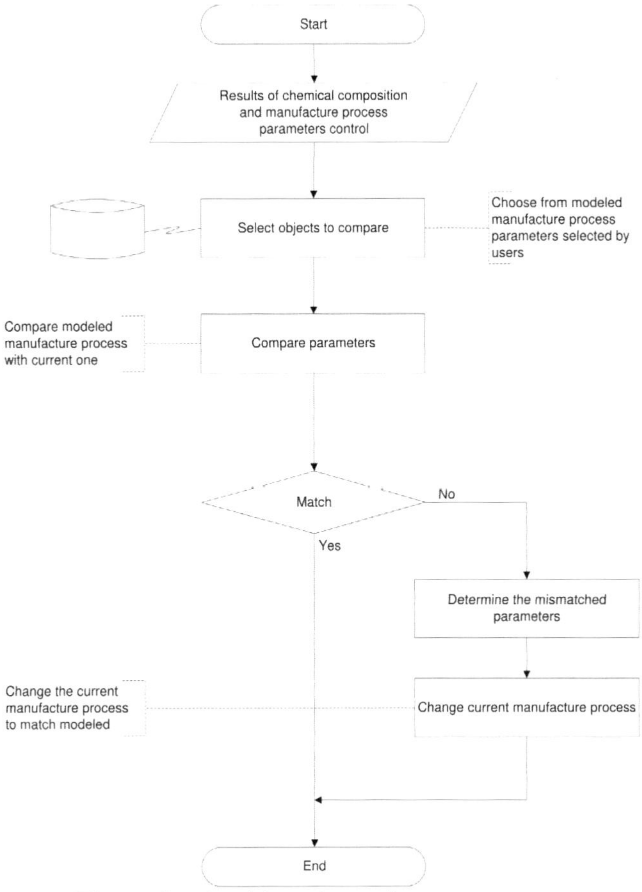

Figure 2. Algorithm of TP adjustment process

consideration the alternatives.

Since at certain stages of the TP as a result of control measurements is formed different amounts of data, we must use more appropriate ways of forecasting. With a small amount of data it is appropriate to use a neural network for forecasting.

If increasing the amount of data more adequate results will be obtained by using predictive models.

3. Conclusion

In this paper we propose the following solution for the implementation of TP control:

1. TP control stage is divided into three parts: the selection of parameters for comparison, the comparison with the simulation results, the change in the current process. First, we define the list of parameters with which the control in the production is carried out. After determining the parameters the pairs are formed for comparison, consisting of values of TP parameters in the production and the values of the simulated TP.
2. TP and chemical composition adjustment is produced by introducing additional elements (elementwise) or modifiers. If adjustment is required only for one element it is used rule base. If it is a mismatching of several elements, modeling will be conducted by forecasting using cascade neural network.
3. TP changing is carried out by means of the rule base, the genetic algorithm and the neural network. With the help of the rule base possible actions by which the TP parameters in the production will match the simulated process are identified. Genetic algorithm based on abovementioned rules, allows you to choose the best way to solve this problem, taking into account the possible combinations of TP parameters.

References

[1] Klochkova KV, P. S., Yusupov LR Formation of the model of knowledge base of intelligent system to design compacted graphite iron // Student scientific journal " Grany nauki" Volume 2, Number 2. The journal is published under the supervision of FGAEI HPE KFU, 2014. p.73-78

[2] Klochkova K., S. Petrovich, V. Abramov "Training of a neural network using genetic algorithm to predict the properties of compacted graphite iron." Student scientific journal "Grany nauki", Volume 1, Number 2 / - Kazan: Publishing house "Kazan (Volga) Federal University" 2013. p. 104-107

[3] Popov AA Application of genetic algorithms to solve optimization problems in the economy. 3 M .: 2009. p.

[4] Internet resource: The method of controlling the process of production of gray and ductile cast iron with nodular and compacted graphite iron for castings. RF Patent 2337973. http://www.findpatent.ru/patent/233/2337973.html (Date of access: 07.08.2014).

[5] Panov AG, Kornienko AE, Degtyareva NG, Zinoviev YA Homogenizing modification of cast iron foundry melts of graphitized cast iron: monograph / AG Gentry [et al.]; The Nizhny Novgorod. state. tehn. Univ. after. RE Alexeev. - Nizhny Novgorod, 2013. 191p.

[6] VV Andreev Features of formation of ductile iron cast structure and the development of effective technologies for the production of castings with high performance operational properties.// Dissertation for the degree of Doctor of Science 4 M .: 2012. p.

[7] S. Osovsky Neural networks for information processing. M.: Finance and Statistics. 2004. 155 p.

[8] Klochkova KV, Shafigullin LN "Analysis of the methods of processing of experimental data of CGI castings." International SPC "Information Technology. Automation. Updating and problem-solving training of highly qualified personnel (PAI 2012) ", December 7-8, 2012 .: collection of works / - Naberezhnye Chelny: Publishing House of the branch FGAEI HPE" Kazan (Volga) Federal University ", 2012. p. 44-49

[9] Russell SD Artificial Intelligence. The modern approach.// Moscow, 2007. P.341-347.

[10] Bloody AD Application of Artificial Intelligence in the metallurgical industry. - Metallurg: TsSRIchermet №7. 2003.

Stages of vermicular cast iron properties modeling in the intelligent design system

K V Klochkova, S V Petrovich, L A Simonova, L R Yusupov
Kazan Federal University, Naberezhnye Chelny Institute
423810, Naberezhnye Chelny, Mira av., 68/19, Russia.

E-mail: kvsimonova@mail.ru

Abstract. This article presents the structure of intelligent system of the cast iron with vermicular graphite iron (CGI) design under the conditions of current production, the technique of the optimal process TP parameters of the production of CGI parts in the preparatory phase of production based on mental models is designed.

1. Introduction

The object of the study is compacted graphite iron (CGI), because at the moment it is the most popular and promising material in the manufacture of parts in mechanical engineering (clutch housing, splitter gear crankcase, cylinder block, transmission housing).

Analysing the source [1] and [2] it can be concluded that in the production of compacted graphite iron is very important to control accurately the chemical composition of the modified cast iron, processes of cooling, casting, primary and secondary modifications.

Intelligent system for predicting the properties of cast iron with vermicular graphite will reduce the complexity of experimental studies, improve a prediction of physical-mechanical and technological properties of materials, which ultimately affect the quality and cost of manufactured products.

Despite the existing designs an urgent problem remains: the development of software modules to predict the chemical composition and properties of cast iron in the existing production. Operative forecast is designed to detect unwanted trends and developing of corrective actions aimed at their prevention.

Today, there are a number of functional and popular systems for forecasting: MedeA, Ansys, ProCAST, SIMULIAAbaqus, and PoligonSoft. However, these systems have high demands to the machine resources and for the prediction a highly qualified staff is required, since it is necessary to interpret the results [3].

The proposed solution is to apply the elements of artificial intelligence, self-organizing neural networks and a combination of methods of mathematical analysis, which will increase the rate of reaction of the system to change the source data. The use of popular software system will reduce the cost and simplify the requirements for the machine resources. As a model of knowledge representation are chosen production rules in the form of the structure of fuzzy sets, as this view helps to solve problems where knowledge meets ambiguity of knowledge (interpretations) [4]. Intelligent system design is based on a unified knowledge base, which consists of the precedent base, rule base and the database [5]. The knowledge base is formed during the technological process at the stages of system functioning. The need to train the system, ie updating the knowledge base, as well as in some cases the lack of ready-made solution creates the need for new experimental studies. To do this, you must make rational planning of experiments with their subsequent statistical processing to design the predictive models. The need for using a mathematical model with a combination of methods of rational experiment planning and statistical processing methods of experimental data is due to the fact that the application of these methods separately in the tasks with a large number of variables does not give adequate results, and if they give, the computation is very time-consuming.

The analysis revealed that the neural network is a flexible and powerful tool for forecasting, due to its ability to educate themselves and complete their education. [6] With their help it is possible to solve both direct and inverse problems, and describe any mathematical model. Selecting a genetic algorithm as a tool for optimization is due to the fact that for the solution of the given task is required a systematic approach, effective methods and criteria for assessing the adequacy of the models, which are aimed not only (not so much) for the maximization of the criteria, but also to optimize the relationship.

The TP adjustment is necessary because there is currently a large percentage of deviations, in technology – alloy overmodification, unacceptable temperature indicators and incorrect choice of modifiers to adjust the composition.

At the moment, to formulate the task of obtaining a certain qualitative composition precisely or diagnose a specified quality with modern production methods is not possible.

2. Basic part

In the result of analysis of TP operations of products manufacturing from CGI by sources [7], [8], [9] and identifying the main dependencies, the structure of intelligent system of CGI design is developed (Figure 1), as well as the technique of the optimal TP parameters formation of CGI parts production in the preparatory phase of production based on mental models.

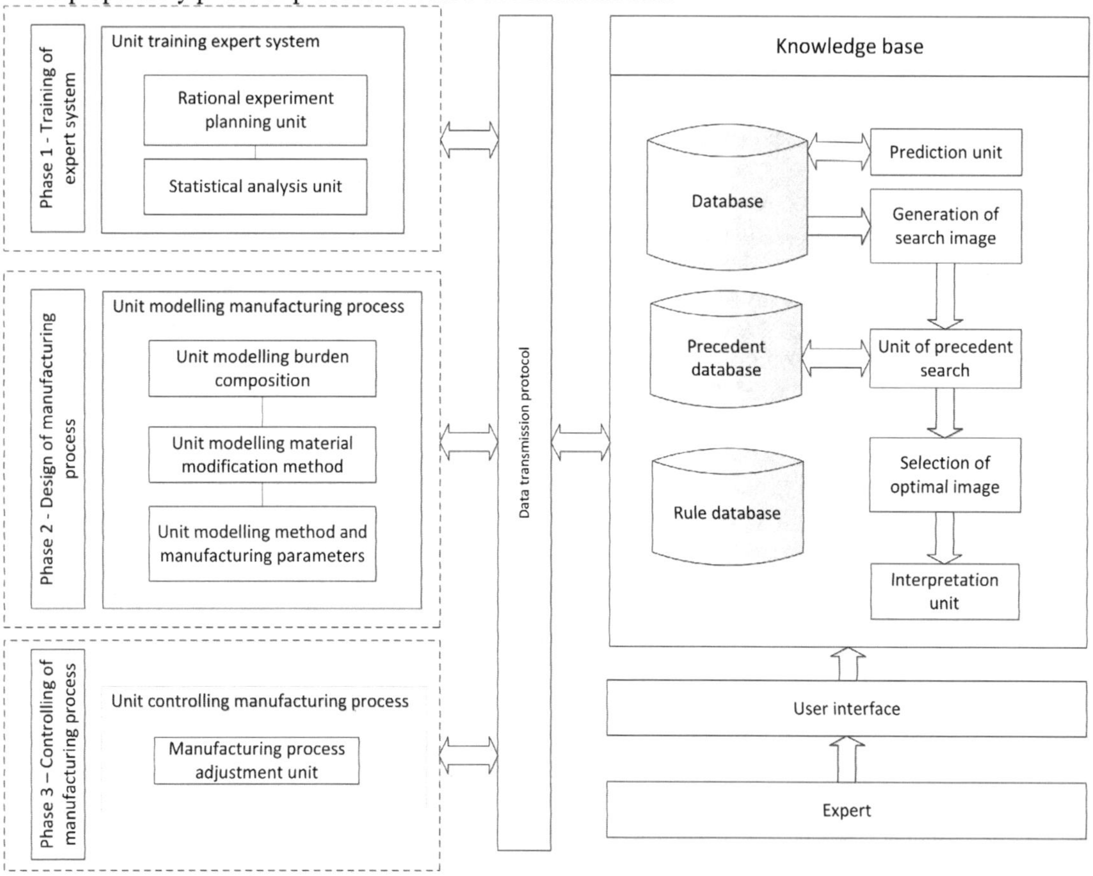

Figure 1 - Structure of the intelligent system of CGI design

Intellectual system functioning of CGI design is divided into three stages: learning stage, the stage of TP modeling, the stage of TP controlling.

The training phase. It is intended to supplement the knowledge base. The training phase consists of the following blocks: block of rational planning of the experiment, block of statistical processing . The need for applying of a mathematical model using a combination of methods of rational experiment planning (by the method of combinational squares) and methods of experimental data statistical processing (using regression analysis and statistical processing block of combinational squares method) is due to the fact that the application of these methods separately in the tasks with a large number of variables does not give adequate results, and if they give, the computation is very time-consuming.

Block of rational experiment planning is for pre-planning of experiments by the combinational squares method, which will reduce their number in the n^{m-2} times (m - number of factors, n- number of factor combinations), while retaining the ability to predict results for the full amount of experiments.

Statistical processing block is designed for experimental data processing with a combinational squares method, and using regression analysis for the predictive models obtaining.

After the experiments the obtained data are recorded in the database, and the parameters of the experiments are recorded as precedents.

TP modeling stage. At this stage, TP operational parameters modeling is carried out, the raw materials (stock) and the finished castings properties. It consists of a stock composition simulation block, modeling block of modifying method, modeling block of method and processing parameters.

Block of stock composition modeling. Prediction of the optimal quantitative chemical elements correlation of the stock on the assumption of the information entered by the user without the influence of factors such as the number and sequence of modifiers input, methods and heat treatment parameters. Simulation is performed using a cascade neural network [10]. To do this, apply a genetic algorithm.Прогнозирование оптимального количественного соотношения химических элементов шихты исходя из информации введенной пользователем без учета влияния таких факторов, как: количество и очередность ввода модификаторов, способ и параметров термической обработки. Моделирование выполняется с помощью каскадной нейронной сети[10]. Depending on the concentration of chemical elements there can be various embodiments of the materials properties, whereby there is a need to optimize the task. To do this, we apply a genetic algorithm.

Block of method of modifying modeling. Depending on the initial chemical composition it is necessary to determine how and in what order to add modifiers to obtain the desired intermediate properties of the alloy, so as with the same chemical composition of the main elements the type of added modifiers can significantly affect the final result. Determination of the composition of modifiers and their percent concentration are determined by searching in the precedents database and by prediction based on a cascaded neural network if no precedent has been found.

Block of modeling of method and processing parameters. Methods and processing parameters at various stages (melting process, exposure process, a pouring into ladle and a mold, cooling and crystallization, defining a casting method, a method of supplying the metal in the mold, and heat treatment) are simulated. Is performed by applying a cascade neural network.

Stage of technological process control (TP). Necessary in case of a mismatch of intermediate values obtained in the simulation operations TP and testing results in the production, in this case of an adjustment TP and chemical composition by introducing additional elements (elementwise) or modifiers. If adjustment is required only for one element is used a rule base. If there is a mismatching of several elements, modeling is conducted by forecasting using cascade neural network.

Exchange of information with intelligent system modules is done through data transfer protocol.

Since the intelligent system consists of stages (modules), it is possible to add or exclude certain modules, depending on the complexity of the problem, which extends the range of application of the intelligent system.

The knowledge base is to create a single data space required for the exchange of information throughout the intelligent system. Knowledge base contains not only a factual information, but also the rules of inference, allowing automatic conclusions of newly introduced facts and, as a result, meaningful information processing.

The knowledge base includes a database, precedents database, rule base. The functional part of the knowledge base is represented by the forming image block to search for a precedent, the search precedents block, the optimum selection block, interpretation block and the prediction block.

At the request of the operator database management system generates the required input data of the local task.

These databases are used to search for options of data with similar characteristics. If the case is not found, the original data transferred to the prediction module. The structure of the precedents base is determined by the structure of precedent and information used in the process. Precedent is an array of data that describes the existing facilities and processes.

Precedent is formed as follows:

$$П_i = S_j, M, V, L, G, I_м, Q, T_п, T_о, HB, \sigma_в, D$$

Wherein $П_i$ – precedent, with the number of elements I; S_j – chemical composition, with the number of elements; M – mass; V – type of molding; L – walls thickness; G – modification method; $I_м$ – modifier; Q – casting method; $T_п$ – melting temperature; $T_о$ – cooling temperature; HB – hardness; $\sigma_в$ – long-term strength; D – mark of the obtained cast iron.

Precedent example:

П1=S1(3.84); S2(0.012); S3(2.44); S4(0.36); S5(0.06); M(12,2); M(74); V (interaxle differential crankcase); L(50); G(Inmold); Iм(CompactMag); Q (in sand molds); Tп(1600); To(1125); HB(180); σв(400);D(ЧВГ45).

Operation of prediction module is based on a usage of the cascade (hybrid) of a neural network.

In the basis of the rules are stored the rules necessary for the functioning and connection of all elements of the intellectual system. As the basis for the creation of the rules is taken the method of fuzzy logic.

Rule example for finding the optimal way of solving on the found similar precedents:

Rule 1

IF «Pi» ≤ «Pmax» and «Pi» = «Pmin» And "Pc" = "High" SHOULD BE "Send to the interpretation block»

Forming the image for search Depending on the entering data and the task under consideration (finding the optimal composition, predicting the properties. TP formation, etc.) there is a request to the database for the necessary information.

The appropriate image of solution is formed, which is an search object in precedent database.

Search of a similar image in the precedents base. If a search of the identical images with a composed solution manner in the precedents base returned no results, it is necessary to search for similar precedents.

Selection of the optimal image. If the system managed to find a few images to solve this task, you need to select the most optimal one. Optimality criteria can be indicators of quality, cost, and the degree of similarity of the found solution.

In case of not finding identical or similar images in the precedents database it is provided a system application to *the prediction block.*

Interpretation block. It is intended to explain the solution found in the form of a precedent for subsequent output to an expert interface.

3. Conclusion

The paper presents an intelligent system of design based on a common knowledge base, which consists of the precedent base, rule base and the database. The functional part of the knowledge base is represented by the image forming block, search precedents block, optimum image selection block, interpretation block and the prediction block. As a model of knowledge representation are chosen production rules in the form of the structure of fuzzy sets. Functioning of the intelligent system of CGI design is divided into three stages: learning stage modeling process, the stage of controlling the process. Determination of the composition of modifiers and their percent concentration is carried out by searching in the precedents database and by prediction based on a cascaded neural network if no precedent has been found. Module operation is based on a prediction of the cascade (hybrid) neural network.

References

[1] Panov AG Getting defect-free structure of cast iron castings using modifiers of the new generation: the theoretical foundations and practical solutions. // Thesis for the degree of Doctor of Science Nizhny Novgorod, 2014. 207p.

[2] Panov AG, Kornienko AE, Degtyareva NG, Zinoviev YA Homogenizing modification of graphitic cast iron melts: monograph / AG Gentry [et al.]; The Nizhny Novgorod. state. tehn. Univ after. RE Alexeev. - Nizhny Novgorod, 2013. p.57-73.

[3] LR Yusupov The study of information systems for forecasting properties of materials. Supervisor - Professor Simonova LA // of Kazan Federal University 2014: Digest of articles / Min of ed. and science; Kazan (Volga) Federal Univ. - Kazan: Kazan.un-t. 2014. p. 284-287.

[4] Klochkova KV, Yusupov LR Analysis of models of knowledge representation to create intellectual system. // Information Technology. Automation. Updating and problem-solving of training of highly qualified personnel (ETAP 2014) ": the international scientific-practical conference. (2014; Naberezhnye Chelny). International Scientific and Practical Conference "Information Technologies. Automation. Updating and problem-solving training of highly qualified personnel (ETAP 2014), "2014 .: collection of works / ed.coll. Simonova LA [Et al.]; ed. Simonova LA, Savitsky SK - Naberezhnye Chelny: Publishing House of Naberezhnye Chelny Institute (branch) FGEI HPE "Kazan (Volga) Federal University." 2014. p.137-144.

[5] Klochkova KV, Petrovich S V., Yusupov LR Formation of the model of knowledge base of the intelligent system for the designing of cast iron with vermicular graphite. // Student scientific journal "Grany nauki" Volume 2, Number 2. The journal is published under the supervision of FGAEI HPE KFU, 2014. p.73-78.

[6] Klochkova K., S. Petrovich, SV. Abramov V V Training of a neural network using a genetic algorithm to predict the properties of compacted graphite iron. // Student scientific journal "Grany nauki", Volume 1, Number 2 / - Kazan: Publishing house "Kazan (Volga) Federal University" 2013. p. 104-107

[7] Internet resources: efficient method of producing pig iron with vermicular graphite iron (CGI) http://www.ruscastings.ru/files/file54.pdf (Date of access: 04.03.2014).

[8] S. Osovsky Neural networks for information processing. M .: Finansy i statistika.2004. 155 p.

[9] VV Andreev, SP Korolev, Panfilov EV New efficient technologies for production of castings made of cast iron with vermicular graphite // Liteinoe proizvodstvo. - 2009. - N 12. p. 5-7

[10] SP Korolev, Panfilov EV, Gurtovoiy DA, Haris S. Korolev MS Management of the production process of castings made of cast iron with vermicular graphite // // Liteinoe proizvodstvo. - 2009. - N 12. p. 13-15

IOP Conf. Series: Materials Science and Engineering **86** (2015) 012017 doi:10.1088/1757-899X/86/1/012017

Development of structural element precedent of technological process in computer-aided design

L A Simonova[1], **E I Egorova**[2]

[1]Kazan Federal Unuversity, Naberezhnye Chelny Institute
423810, Naberezhnye Chelny, Mira av., 68, Russia.
[2]Kazan National Research Technical University named after A.N.Tupolev - KAI (KNRTU-KAI), Russia

E-mail: chelny@kpfu.ru

Abstract. Based on the methodology of building of the intelligent system of TP design, a new approach to the transformation of information on structural and technological characteristics of the part and of 3D part model and representation of it in the form of elementary constructive modules structural units of a surface to form automatically alternative TP variants on a complex of qualitative and quantitative criteria. Each constructive surface module is represented in the database of module CAM of the intelligent system for constructing TP with records of all possible transitions.

1. Introduction

A large number of unresolved problems are related to the automation of the design of technological processes (TP). Almost all domestic CAM TP (CAM (SPP- system of production preparation) systems) do not allow today to automate the TP design of a part manufacturing. In modern CAM TP used mainly the designing of routing technological processes based on analogues (standard, TP group, parameterized models of TP, "common" TP for a group of parts) [1] .Methodically different approaches are used: level view of technological processes, the representation of processes as "trees", etc. The role of technologist-designer is crucial so as he forms the routing technological process, based on his own knowledge, experience, intuition and preferences (often erroneous). Design decision is subjective.

Meanwhile, the technological process - is, first of all, the routing technological process and the accompanying it additional information about the place of his realization, used equipment, the expected labor costs. The designed technological process is the carrier of information used by the various subdivisions of the enterprise to manage current production, analysis and prediction of the new one [1, 2].

A truly creative is precisely the formation of routing TP and defining the means of technological equipment. The rest is derived from this process. However, at this important stage of designing the existing CAM TP practically do not give the technology the necessary intellectual support. All subsequent steps of the designing are less complex, but are associated with a significant amount of routine work - preparation of technical documentation, various statements and specifications. These functions in modern CAM TP automate successfully. [2-6]

It could argued that the vast majority of existing CAM TP (both of domestic and foreign design) are automation systems of technological operation level. These systems allow to raise productivity of a technologist through automation of routine work related to the process of designing, ordering relationships of the designers in operation process, providing a wide range of convenient service functions of conducting automated archives etc. These factors contribute to improve the quality of technologist's labour as well as they streamline their work and allow to focus more attention on the adoption of the effective technological solutions.

However, the potential for improving the quality of the design of technological solutions - the formation of their high-performance structures, is currently unused in creating intelligent CAM TP.

Similarities and luck of originality of scientific and methodological approaches used in designing of systems make some CAM TP similar in features and little distinguishable that significantly reduces their competitiveness [1].

2. Basic part

Analysis of the nomenclature of parts of engineering production showed that in the process of their production cycle the mechanical treatment is dominant [2]. In papers [2, 3] the methodology of TP construction on the design and technological characteristics of the part in the intelligent system of TP construction is described. 3D part model is represented as an elementary constructive surface modules. Each constructive module of the surface is represented in the database of CAM module of the intelligent system for constructing TP by records of all possible transitions (Figure 1).

When creating intelligent systems frame-based models in which concepts are linked with each other by different relationships are used. Relationship "is - some« (isa) » allows to organize the structure of concepts in the form of a tree. In the engineering interpretation functions of concepts standardized or uniformed largely. Their descriptions are contained in the extensive glossary of terms. Concepts (Table 1) are related by generic relationships. The term "product" is the generic and its variants - the concepts of "assembly unit" and "part".

With the help of a relation a complex concept can be separated into components by decomposition. Constructing TP is necessary to ensure the selection of the structural element at a certain stage on the qualitative and quantitative indicators.

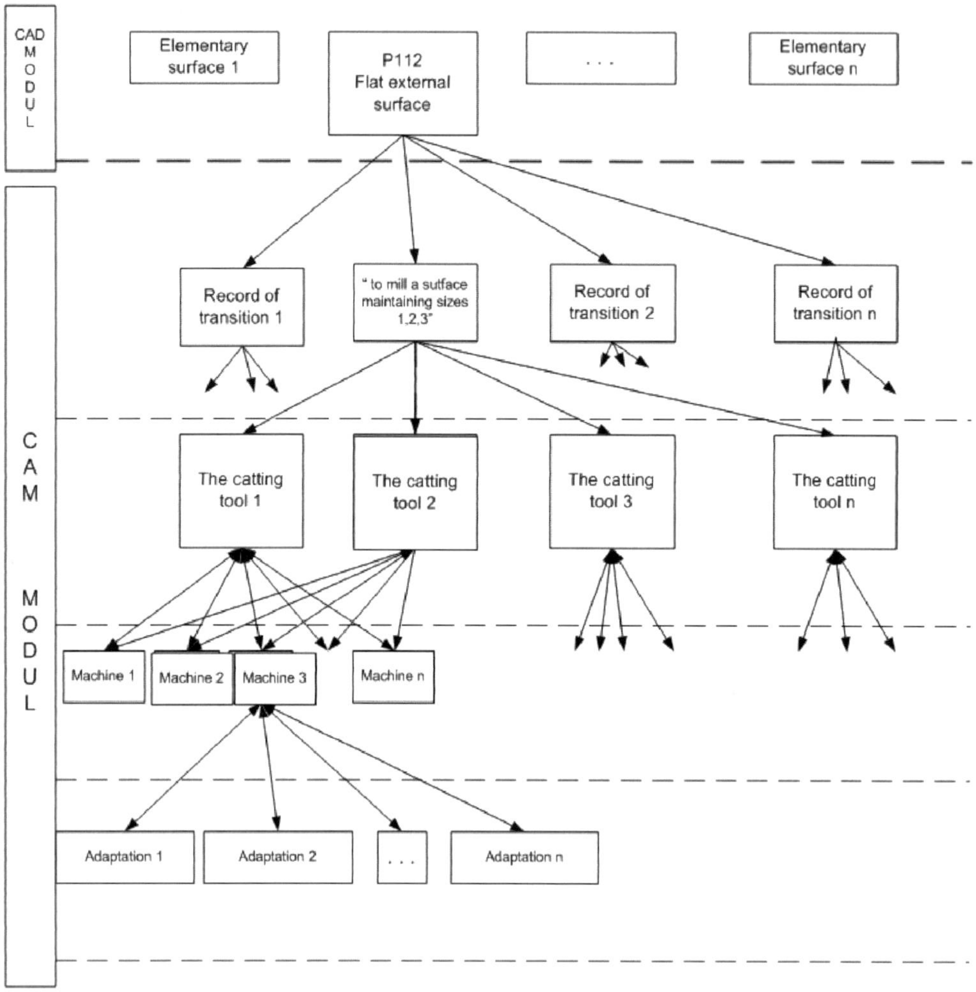

Figure 1. Levels of TP differentiation

Qualitative indicator is the complex value, which will depend on the quality indicators of each structural element.

The important quality characteristics of the TP are the processing time, which the company aims to reduce and the productivity of the process that the company aims to increase. Levels of the hierarchy of the TP, such as OPER, POZ and SET are characterized by the selected metal-cutting equipment. In modern engineering is advisable to aim to apply modern machining centers with CNC, where multi-processing which will reduce the install / reinstall is possible. This would reduce the normal time processing and therefore enable to increase a production. At levels of TP hierarchy, such as TRANZ and PS (power stroke) the quality indicator will be machining stock, which is a calculated value. Technological problem is solving, is it possible to remove machining stock for one PS or not. If the machining stock is less than the calculated value, the quality of a specified by a drawing will not be obtained in the course of treatment, if machining stock is higher than the calculated values, the requirements for a specified quality will be done, but at great costs, that is, the production of the process will be reduced, and the costs of the processing will increase. Therefore it is necessary to seek to ensure that the removed machining stock will be equal to the calculated value.

Quantitative parameters of the TP are all the values that are calculated or taken from a drawing. It is very important at the design stage to eliminate calculation errors and make a choice of machines, tools, appliances not "approximately", but more or less matching the calculated values. Of course, in a rapidly developing world, you must always be aware of the timing of technological preparation of production. According to [2] with the general ideas in mechanical engineering, TP is implemented in seven hierarchical levels, which are allocated as a result of systemic differentiation of its functions.

From the definition of the level of decomposition of the transition (TRANZ) - transition is characterized by constancy of used cutting tools. The function of the equipment is a list of ongoing technological methods that is common to the cutting tool and equipment. Operation (OPER) is characterized by the constancy of the workplace, and the setting (SET) by the constancy of basing conditions and fixing of the workpiece in the device. Device function is mounting and fixation of the workpiece and enforcement of the operation. Thus, the interaction of design objects through their function on the higher level of decomposition is carried out (Figure 1).

To each " transfer record " of a particular structural module is associated a group of cutting tools. The cutting tool type may be more than one for one transition record. Selecting the type of cutting tool depends on the structural and technological parameters of parts, machine model, the type of production and is based on the certain rules. Databases of " transition records " are associated with the database of the group and the type of cutting tool by the defined rules. The function of the cutting tool is the implemented technological method.

For the automatic selection of the cutting machine, cutting tools, machine retaining device it is necessary to design an automated selection module.

Selection of the machine, tools, device occurs in the course of the design process, and therefore it is necessary to use in a data domain a number of properties proven in practice, the structural elements of the processes - analogues. Input information for the selection of the cutting machine at the operation level is the TP indicator. Input information for the selection of the cutting tool at a transition level is TP indicator. Input information to select the device on the position level is the TP indicator. Thus information of the object and its structural components is stored in the database. Great importance for the choice of cutting tools, machine and device is the content of the transition (transient recording), its parameters, because at this level a frame model for the position, operation, a part of which is this transition, is constructed.

Base of precedents are formed on the basis of the rules of choice of cutting tools, machine and device. Precedents are the pairs like "operation - the machine", "operation - a tool", "operation - a device"; "Position - the machine", "position - a tool", "position - a device"; "Transition - the machine", "transition - a tool", "transition – a device"; "Process - the machine", "process - a tool", "process - a device."

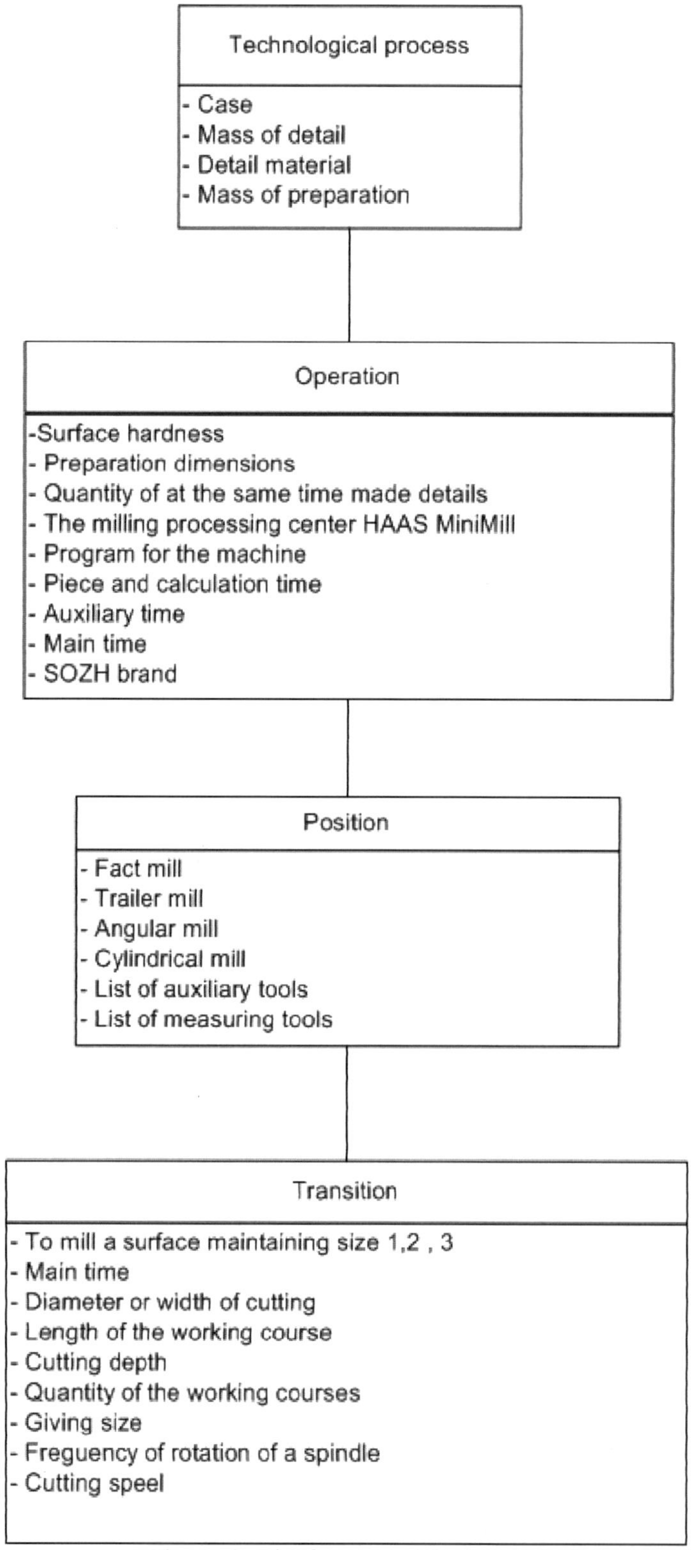

Figure 2. The frame model of the TP

IOP Conf. Series: Materials Science and Engineering **86** (2015) 012017 doi:10.1088/1757-899X/86/1/012017

In the module CAM of intelligent system of TP construction is suggested an idea of the TP as a hierarchy structure of the related classes and objects.

Constructive module of the part surface is presented in a CAM module by the transient recording.

Transient recording is associated with the cutting tool, device and cutting machine by the frame model.

The appropriate part of this stage in a single database system of intellectual TP construction system is designed.

3. 3. References

[1] CAM processes: a textbook for stud. of high school / AI. Kondakov.-M.: Publishing center "Academy", 2007.-272p.

[2] Information support of technological routes. Integrated information systems in mechanical engineering / L. Simonova. - Palmarium Academic Publishing GmbH & Co.KG, 2012-186 p.

[3] Intelligent system of technological processes designing on structural and technological characteristics of the parts / EI Egorova, V. Matveev - Forging - stamping production. Materials processing pressure. 2012. № 11. p. 33-40.

[4] Simonova LA, Kostyuk IV 2009 Formation of knowledge base system of tooling management Kuznechno-Shtampovochnoe Proizvodstvo (Obrabotka Metallov Davleniem) № 8 pp 22-28

[5] Simonova LA Formation of knowledge database of alternative technological routes' parameters (2004) Kuznechno-Shtampovochnoe Proizvodstvo (Obrabotka Metallov Davleniem) (10) pp 44-48

[6] Larisa A Simonova, Marat R Khisamutdinov 2013 Intellectual Model Control Data of the Module Integration SAP-ERP and Team center-PLM. World Applied Sciences Journal 25 (8): pp 1258-1262

[7] AV Zatonskii Synthesis of expert systems of management of socio-technical systems // Mountain information-analytical bulletin (scientific and technical journal) = Mining informational and analytical bulletin (scientific and technical journal). -2008. -№ 2. P. 82-86

[8] Norenkov IP, Kuzmik PK 2002 Information support of high-tech products CALS-technologies - M .: Publishing house MSTU. N E Bauman 320 p

[9] Knowledge Base of Intelligent Systems / TA Gavrilova, VF Horoshevsky - St. Peter 2000 384 p Il

[10] Intelligent systems of design: tutorial / GB Eugenev Moscow: Publishing House of the MSTU. N E Bauman 2009 334 [2] p Il

IOP Conf. Series: Materials Science and Engineering **86** (2015) 012018 doi:10.1088/1757-899X/86/1/012018

Automated test system of internal combustion engines

L A Galiullin

Kazan Federal University, Naberezhnye Chelny Institute
423810, Naberezhnye Chelny, Mira av., 68/19, Russia.

E-mail:galilenar@yandex.ru

Abstract. The method of engine test bed control based on fuzzy neural network is considered. The structure of fuzzy neural network to control the engine during the test is suggested.

1. Introduction

Leading engine company conducts intensive research and development work to improve the reliability and durability of the internal combustion engine (ICE), and in particular diesel engines [1].

Research and testing of diesel engines are one of the main means of checking the quality of the manufacture of parts and assemblies, sub-assemblies and engine as a whole, its proper mounting, compliance of the essential characteristics of a diesel engine with the requirements of technical documentation [2].

Types of diesel tests are regulated by GOST and international standards ISO, which govern the the rules of an acceptance and requirements to the technical level of engines. After acceptance and putting diesel engines into production an improvement of their designs and technical and economic indicators continues.

Currently, testing of diesel engines is a complex and time-consuming technological process, very similar to the pilot study. Therefore, an engine automated test system (ASI) has been created.

Modern requirements to continual improvement of the technical level of manufactured diesel engines lead to the fact that the share of the costs of the tests in the process of creation of new engine models grows more and more. Particularly large, these costs are in the case of non-compliance of the level of production automation and scientific research one. In this regard, test technological process automation is one of the main objectives of improving the technological level of production and quality of the diesel engines.

Normalization of the input parameters of a diesel engine

AST should perform optimal control diesel engine during the tests at established modes, ensuring at any time the required values of engine output parameters.

To do this, AST on the basis of forming fuzzy neural network knowledge base generates control action [3]. Control action for diesel is moving the high pressure fuel pump (HPFP), h regulator.

The knowledge base has the form of fuzzy control rules. Control rules are constructed using the theory of fuzzy sets and fuzzy logic:

$$R^{(k)}\text{: IF } \omega \text{ AND } M_H \text{ AND } G_T \text{ AND } \dots \text{ THEN } h, \qquad (1)$$

wherein k – the total number of fuzzy rules;
$R^{(k)}$ – the entire set of rules;
ω – linguistic variable characterizing the engine speed;
M_H – linguistic variable characterizing the torque load;

G_T – linguistic variable characterizing the fuel consumption;
h – linguistic variable characterizing the position of the injection pump regulator.

Fuzzy rules are clear and simple, in contrast to the differential equations describing the engine and its system [4].

Mathematical algorithm of AST diesel work provides preliminary input of parameters of the engine. Let the input parameters for AST in the test set: crankshaft speed – ω (min^{-1}), torque – M_H (Нм), hourly fuel consumption – G_T (kg/hour). However, it is possible to set a greater number of input parameters.

Next is the normalization of each input parameter.

1-st step – calculation of the arithmetic mean of the parameter:

$$\bar{x} = \frac{\sum_{i=1}^{n} x_i}{n} \tag{2}$$

wherein x_i – parameter value;
n – the number of observations.
2-nd step – calculation of standard deviation:

$$\sigma = \sqrt{\frac{\sum_{i=1}^{n}(x_i - \bar{x})}{n-1}} \tag{3}$$

3-d step – translation values of the points in the 10-point scale :

$$St_i = \frac{(x_i - \bar{x})}{\sigma} \times 2 + 5{,}5 \tag{4}$$

2. Fuzzification of normalized engine parameters

Determination of membership level of normalized parameter of the engine to the given functions (fuzzification) is carried out using standard Gaussian function represented in a rational form:

$$\mu_A(x_i) = \frac{1}{1 + \left(\frac{x_i - c_i}{\sigma_i}\right)^{2b_i}} \tag{5}$$

wherein x_i – the normalized value of an indicator;

c_i – parameter of center of formal neuron set on the interval [0, 10] points of 10-point scale. Depending on the given number of formal neuron the 10-point scale is divided into corresponding number of sections, where the parameter c_i – midpoint of the segment;

σ_i – parameter (ratio) of function latitude. Originally is generated by the automated system is equal to 2/3 of the segment setting by the parameter c_i.

b_i – shape parameter of function. Originally is generated by an automated system to 1, which corresponds to the standard Gaussian function.

AST conducts fuzzification for each vector of the input parameters of the engine.

3. Formation of fuzzy inference rules for the knowledge base of AST

For each of the formal neurons is determined the integral degree of membership of all parameters of the engine introduced into AST. Для каждого из формальных нейронов определяется интегральная степень принадлежности всех введенных в АСИ параметров двигателя.

It is used in the AST the degree membership aggregation of individual parameters using the procedure of the algebraic product, from which it follows that for the k-th rule of inference::

$$\mu_A^{(k)}(x) = \prod_{j=1}^{N} \left[\frac{1}{1 + \left(\frac{x_i - c_j^{(k)}}{\sigma_j^{(k)}} \right)^{2b_j^{(k)}}} \right] \tag{6}$$

4.Calculation of control actin to the diesel

Approximation of fuzzy sets into the exact solution h is conducted by the system using a model of Mamdani-Zade.

When M inference rules and the use of the generalized Gaussian function as membership function, moving the high pressure pump regulator is determined by the formula:

$$h(X) = \frac{\sum_{i=1}^{M} c_i \left[\prod_{j=1}^{N} \left[\frac{1}{1 + \left(\frac{x_i - c_j^{(k)}}{\sigma_j^{(k)}} \right)^{2b_j^{(k)}}} \right] \right]}{\sum_{i=1}^{M} \left[\prod_{j=1}^{N} \left[\frac{1}{1 + \left(\frac{x_i - c_j^{(k)}}{\sigma_j^{(k)}} \right)^{2b_j^{(k)}}} \right] \right]} \tag{7}$$

wherein $c_j^{(k)}$, $\sigma_j^{(k)}$, $b_j^{(k)}$ determine center parameters, width and shape of the j-th component x to the k-th fuzzy inference rule.

The structure of fuzzy neural network to diesel engines [5] AST, which implements the approximation (7) function is shown in Figure 1.

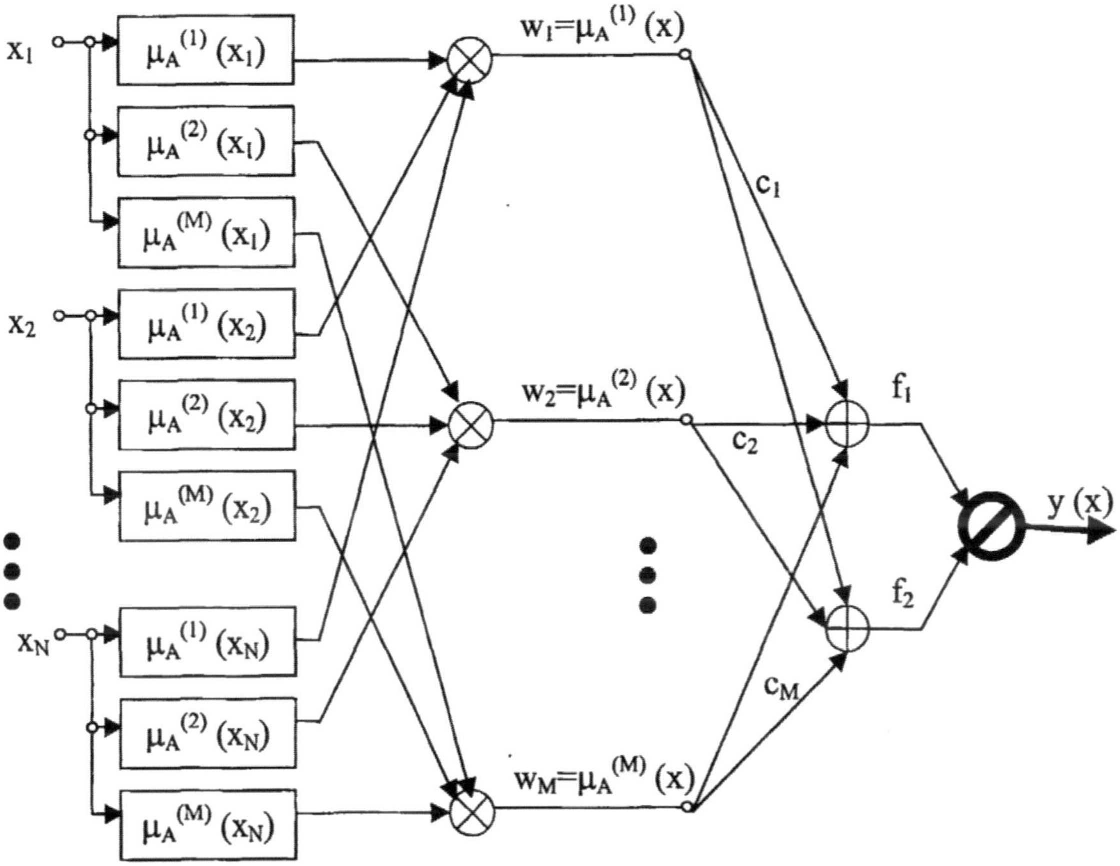

Figure 1. The structure of the fuzzy neural network diesel AST

This four-layer structure in which the first layer performs separate fuzzification of each of the input variables x_i (i=l, 2, ..., N), determining for every k-th inference rule coefficient of belonging value $\mu_A^{(k)}(x_i)$ [6].

The second layer performs the aggregation of individual variables x_i, determining resultant coefficient of belonging value $w_k = \mu_A^{(k)}(x)$.

The third layer – aggregation of M inference rule (the first neuron – f_1) and generation of normalizing signal (the second neuron – f_2):

$$f_1 = \sum_{i=1}^{M} c_i \left[\prod_{j=1}^{N} \left[\frac{1}{1 + \left(\frac{x_i - c_j^{(k)}}{\sigma_j^{(k)}} \right)^{2b_j^{(k)}}} \right] \right],$$

$$f_2 = \sum_{i=1}^{M}\left[\prod_{j=1}^{N}\left[\frac{1}{1+\left(\frac{x_i - c_j^{(k)}}{\sigma_j^{(k)}}\right)^{2b_j^{(k)}}}\right]\right] \tag{8}$$

A fourth layer consisting of one neuron generates an output signal $y(x)$.

5.Conclusion

The proposed methods and algorithms are implemented in the form of a computer program "Automated Information System "Testing diesel internal combustion engine based on fuzzy neural network"".

1. When AST works with a first pair of data $(y_1(x), d_1)$ is created the first cluster with the center $c_1=y_1(x)$. It is taken, that $w_1=d_1$ and that the cardinality of the set L_1=l. Let r denotes the marginal Euclidean distance between $y(x)$ and the center, in which data will be interpreted as belonging to the generated cluster. To preserve the generality of the solution is assumed that at the moment of the training start there are M clusters with the centers $c_1, c_2, ..., c_M$ and the corresponding values w_i и L_i (i=1, 2, ..., M).

2. After reading the k-th training couples $(y_к(x), d_k)$ are calculated the distances between $y_к(x)$ and all existing centers $\|y_к(x)-c_l\|$ for l=1, 2, ..., M. Assume that the nearest center is c_{lk}. In this case, depending on the value $\|y_k(x)-c_l\|$, may appear one of two situations :

- if $\|y_k(x)-c_l\|>r$, it will be created a new cluster $c_{M+1}=y_к(x)$, and $w_{M+1}(k)=d_k$, $L_{M+1}(k)=1$. Parameters of the prior established clusters do not change, i.e. $w_l(k)=w_l(k-l)$, $L_l(k)=L_l(k-l)$ for l=1, 2, ..., M. Number of clusters M is incremented by 1 ($M = M$+1);

- if $\|y_k(x)-c_l\|<r$, data will be included into l_k-th cluster, the parameters of which are specified in accordance with the formulas:

$$w_{lk}(k) = w_{lk}(k-1) + d_k \tag{9}$$

$$L_{lk}(k) = L_{lk}(k-1) + 1 \tag{10}$$

$$c_{lk}(k) = \frac{c_{lk}(k-1) \times L_{lk}(k-1) + y_k(x)}{L_{lk}(k)} \tag{11}$$

3. After verifying the parameters of the fuzzy system, function that approximates the AST diesel input data, is defined as:

$$f(y(x)) = \frac{\sum_{l=1}^{M} w_1(k)\exp\left(-\frac{\|y(x)-c_1(k)\|^2}{\sigma^2}\right)}{\sum_{l=1}^{M} L_1(k)\exp\left(-\frac{\|y(x)-c_1(k)\|^2}{\sigma^2}\right)} \tag{12}$$

whereas other clusters do not change, i.e. at $l \neq l_k$, $w_l(k)=w_l(k-l)$, $L_l(k)=L_l(k-l)$ для l=1, 2, ..., M.

By repeating these steps of the algorithm until $k=p$ clarifying each time values of M data space is divided into M clusters, wherein the power of each of them is defined as $L_l=L_l(к)$, center – as $c_l= c_l(k)$, and the value of the accumulated function attributed to him d – as $w_l=w_l(k)$.

Separation of data space into clusters occurs independently and without the participation of a technologist, in accordance with a predetermined threshold value r . In developed diesel engine AST value of r is set at the established amount of formal neurons:

$$r = \frac{10}{v} \tag{13}$$

wherein 10 –points of 10-point scale;

v – number of formal neurons set by the user of the system.

The effectiveness of the proposed neuro-fuzzy system to control diesel is examined. Since the setup time of the stand is reduced, time savings is 25% , as the result the fuel savings is 17% .

References

[1] Zhou M L, Wang J C and Li Y P 2014 Automobile engine fault diagnosis and prediction system *Advanced Materials Research* **1008** 641-644

[2] Zhou Q, Gullitti A, Xiao J, Huang Y 2008 Neural network-based modeling and optimization for effective vehicle emission testing and engine calibration *Chemical Engineering Communications* **195(6)** 706-720

[3] Zubkov E V and Galiullin L A 2011 Hybrid neural network for the adjustment of fuzzy systems when simulating tests of internal combustion engines *Russian Engineering Research* **31(5)** 439-443

[4] Yao Z T and Pan H X 2014 Engine fault diagnosis based on improved BP neural network with conjugate gradient *Applied Mechanics and Materials* **536** 296-299

[5] Shatnawi Y and Al-Khassaweneh M 2014 Fault diagnosis in internal combustion engines using extension neural network *IEEE Transactions on Industrial Electronics* **№61(3)** 1434-1443

[6] Hao P 2014 An improved back-propagation neural network algorithm *Applied Mechanics and Materials* **556-562** 4586-4590

The research of anodic microdischarges in plasma-electrolyte processing

L.N. Kashapov[1], N.F. Kashapov[1], R.N. Kashapov[2]
[1] Kazan Federal University, 18 Kremlyovskaya St
[2] Kazan Physical -Technical Institute, Sibirsky tract 10/7

E-mail: kashramil.88@mail.ru

Abstract. The article is devoted to the topic of anodic microdischarges in plasma-electrolyte processing. The aim of this work is to research the conditions of anodic micro-discharges during the plasma-electrolytic treatment and the influence they have on the surface of metals. As a result of experimental researches, was made a mechanism of influence anodic microdischarges on the surface of the electrode, burning of anodic microdischarges occur in the voltage range of 40-100 W.

1. Introduction

Recently, the discharge of liquid electrodes has found wide application in chemical engineering, materials processing technologies and in other fields of science and technology [1, 2]. With these types of discharges, there is the possibility of formation of surface micro relief [3, 4]. This is due to the fact that the classical methods of treating the surface of electrical steel have disadvantages [5]. In this regard, an urgent need to develop and research new energy-efficient, environmentally friendly and cost-effective methods of treating the surface, based on the use of liquid discharges from the electrodes. The purpose of this study was to examine or testify the conditions of anodic micro-discharges during the plasma-electrolytic treatment and the influence they have on the surface of metals.

2. Experimental

Was developed and constructed an experimental setup to study the anodic microdischarges, the conditions of occurrence and their impact on the metal surface (fig.1).

The experimental setup operates in the range parameters of the voltage $U = 0 \div 250$ B, current $I = 0 \div 50$ A, pH = 1-12 and T = 20-110 ^{0}C. It consists of a current source 1, the electrolytic cell 2, the electrode system 3, of the oscilloscope 4, additional resistance 5, voltmeter 6, and ammeter 7. Camera installation is made in the form of a "tee", two 10 mm diameter tubes, welded together at right angles to each other. The length of the tube base was 80 mm, perpendicular to the tube 30 mm. The power supply provides a regulated constant voltage supply for the power connection to the electrode system. The anode was made in the form of a rectangular plate of width 5 mm, a length of 8 mm and a thickness of 1 mm. The cathode was made of copper in the form of a ring with a diameter of 8 mm and a width of 5 mm. The camera was mounted on a working microscope stage IIB-2, contained in a horizontal position. Through the camera installation was pumped electrolyte. Then, lens was applied to the wall of the plant and focused on the surface of the plate - the anode with a movable microscope tube. Was carried out observation of occurrence anodic microdischarges and their record on the camcorder Sony DCR-TR V60E.

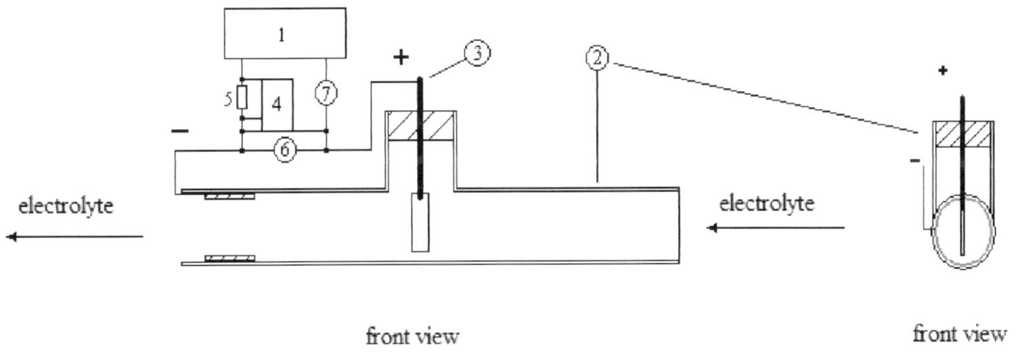

Fig. 1 Functional scheme of the experimental setup, intended to fix the ignition anode microdischarges.

Pumping and purification system is filled with the investigated electrolytes the necessary concentration and composition. In a study of electrolytic cathodes used aqueous solutions of NaCl, Na2CO3, Na2SO4, NaOH, KOH, HCl, H2SO4, concentration up to 15%.
Investigations were performed on specimens made of steel 12X18H9T, steel #3, electrodes E46-ANO-4-d-UD-A (Fe -99%, C -0.1%, Mn-0, 52%, Si - 0,15%), copper and graphite. Samples of the electrode after the plasma electrolytic process were investigated using a scanning electron microscope.

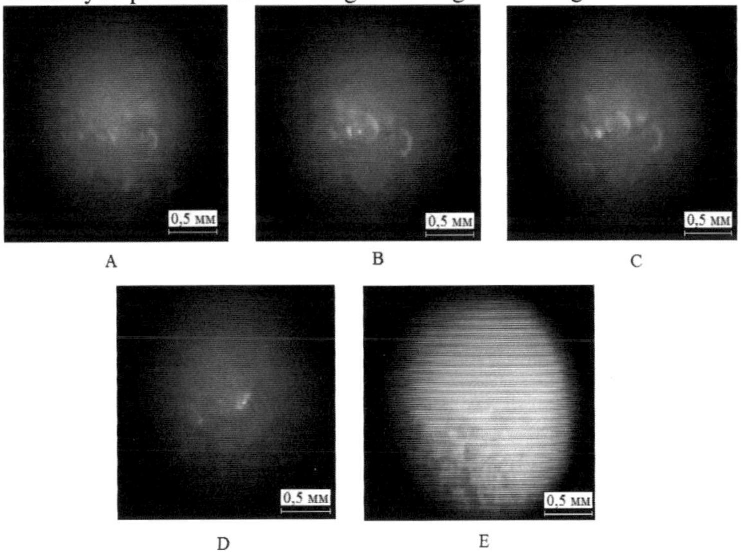

Fig. 2. A series of photos occurrence of anodic microdischarges during plasma electrolytic process.
Were recorded anodic microdischarges in the voltage range from 40 to 90 V. (Figure 2 A - D). Average optical radius of the luminous zone of a single microdischarge was 0.05 mm. These anodic microdischarges are similar to those previously known in /2/ that have the same size. But there are two very strong differences:

1. The material of the anode. In [2], the anode was made of valve metal. The characteristic property of a group of valve metals - forming property in the anodic polarization of a protective oxide layer with high resistance. In our case, it was ordinary steel, the anodic polarization is anodic dissolution of metal.
2. The voltage. In [2], it was 340 V.

A further increase in voltage from 100 to 160 V led to the initiation of a large anode discharge of 0.5 - 1.5 mm, the image of which is shown in Figure 2E.

Based on these studies we propose a mechanism of influence anodic microdischarges on the surface of the electrode. This mechanism involves two action modes of anodic microdischarges on the surface.

The first case - formation of steam jets as a result of evaporation of the electrode. The depth of the hole h $_{hole}$, formed as a result of evaporation, determined by the law of conservation of energy: the latent heat of $q_{el} \cdot \tau_{ex}$ and delivered during the discharge is expended in heating the material to the boiling point, the transition to the gaseous state, the message is a pair of kinetic energy and heat into the interior of the electrode

$$h_{hole} = \frac{\left(q_{el}\tau_{ex} - 2T_{emit}/\sqrt{\pi}\right)\sqrt{\lambda_{\mathrm{T}} c_{\mathrm{T}}\, \rho_{el}\tau_{ex}}}{\rho_{el}\left(c_{\mathrm{T}} T_{emit} + Q_{emit} + 0{,}5 v_{\mathrm{steam}}^2\right)}$$

where v_{steam} - speed of the steam jet, Q_{emit} - heat of phase transformation.

Fast entry of a large amount of heat can cause uneven thermal expansion of the electrode material and create a force breaking away from the solid particles of the electrode. Was observed in the case of plasma-electrolytic formation of the microrelief of graphite electrodes.

The second case - the melting the surface layer of the anode, the melt is continuously removed, and the heat flux acts to ever-exposed solid material of the electrode. The depth of hole was determined assuming that the heat source moves along the boundary of the melt. From the conservation law should:

$$h_{hole} = \frac{q_{el}\tau_{ex}}{\rho_{el}\left[c_{\mathrm{T}}(T_{melt} - T_0) + Q_{emit} + 0{,}5 v_d^2\right]}$$

where v_d - speed of dispersion of the particles melt.

A result of research determined that the burning of anodic microdischarges occur in the voltage range of 40-100 W. A result of research determined that the burning of anodic microdischarges occur in the voltage range of 40-100 W. Based on the analysis of surface morphology of the electrodes and the burning time of anodic microdischarges we propose a mechanism their effect on the electrode surface.

References

[1] Paulmier T, Bell J M, Fredericks P M 2008 *Journal of Materials Processing Technology*, 208, Issues 1-3, 117–123.
[2] Galyautdinov R T, Kasparov N F and Luchkin G S 2002 *Inzhenerno-Fizicheskii Zhurnal* **75** 170–173
[3] Kashapov L, Kashapov N, Kashapov R 2013 *J. Phys.: Conf. Series* **479** (1), 012003
[4] Kashapov L, Kashapov N, Kashapov R 2013 *J. Phys.: Conf. Series* **479** (1), 012011
[5] Kashapov L, Kashapov N, Kashapov R 2013 *J. Phys.: Conf. Series* **479** (1), 012005

IOP Conf. Series: Materials Science and Engineering **86** (2015) 012020 doi:10.1088/1757-899X/86/1/012020

Practical experience of using ultrasound flowmeters at the measurement associated petroleum gas

M G Fazlyyyakhmatov, N F Kashapov, Kh A Khayritonov, D K Lazarev and V K Lazarev
Kazan (Volga Region) Federal University, 18 Kremlyovskaya str., Kazan 420008, Russian Federation

Email: mfazlyjy@kpfu.ru

Abstract. The results of field tests of several ultrasound flowmeters at existing oil and gas extraction objects are given in the paper. Measured medium – associated petroleum gas. This work aims to create a unified system for measuring the amount and parameters of APG in order to reduce operating costs.

1. Introduction

Associated petroleum gas (APG) defined as a gas dissolved in oil, which is extracted from the bowels of the earth with oil. This gas is separated from oil by a multi-stage separation on the oil treating and production objects.

According to [1,2] APG – a mixture of hydrocarbon gases released from crude oil during its production, transportation, treatment.

For a long time the APG was burned directly at the extraction objects. However, in recent years its recycling or safe disposal takes place. It is connected with toughening of requirements in the field of environmental safety as well as the requirements in the field of rational nature management and resource saving. In this connection the problem of accounting the APG becomes relevant.

Accounting for dry natural gas in the world practice was organized long ago. Sets of measuring instruments and methods of accounting approved in practice are used such as commercially produced flowmeters based on different physical principles, as well as calculators with the necessary mathematical apparatus and systems of archiving parameters of the gas.

APG component composition significantly differs from the composition of natural gas. The density of the natural gas varies in the range 0.65-0.7 kg/m^3. The density of the APG is strongly dependent on the component composition and may vary in the range 0.7-1.5 kg/m^3.

The second feature is the physical behavior of the APG because of its complex component composition in the measured pipeline under different conditions.

Accounting for this gas is complicated by the fact that the droplets of hydrocarbons and water, flying in the gas pipeline, are deposited on the sensor of flowmeters, accumulate in the pipe, and the measured medium becomes non single-phase. As a result flowmeter readings are not reliable.

Our research team is working on the selection of the optimal composition measuring instrument [3] for the organization of accounting APG in the oil and gas production objects, according to [1]. This work aims to create a unified system for measuring the amount and parameters of APG in order to reduce operating costs.

Measurement system of amount and parameters APG – a set of functional united measuring instruments, data processing systems and process equipment designed to measure the amount and parameters APG, display and registration of measurement results.

In this paper the results of field studies of several ultrasonic flowmeters from various manufacturers are presented.

2. Measuring Technique

Ultrasonic flowmeters measure the volume of passing gas unlike thermal-mass [3] and Coriolis flowmeters. For accounting gaseous media is used ultrasonic flowmeters using two principles determine the flow rate. First, pulse-time method, the second, the correlation method.

Operating principle of pulse-time flowmeters is based on measurement of the difference between the time of transmission probe pulses in the direction of gas flow rate in the pipeline and the time against him. Emitting/transmitting probe pulses realized like at a most ultrasonic devices [4,5].

Ultrasonic transducers mounted on the measuring pipeline provides emission and reception of probe pulses. Electronic control unit carries out alternately switching emission/reception modes of sensor pairs. Calculator of the flowmeter using the known parameters (propagation time, the inner diameter of the pipeline and the distance between the transducers) calculates the average speed, instantaneous flow rate and generates a normalized frequency output signal proportional to the volumetric flow. The emitting and receiving transducers may be located across from each other - on opposite sides of the pipe, or on one side using reflection from the pipe wall. The advantage of this method – a wide dynamic range.

The sensor of the flowmeters using Doppler method consists of four ultrasonic transducers, arranged in pairs diametrically in two section perpendicular to the axis of the pipeline. The operating principle of the flowmeter sensor is based on measuring the time of movement heterogeneity (small solid particles, condensate droplets) between the two pairs of transducers with a known distance between them. Two emitters radiate ultrasonic oscillations, which, after passing through the gas flow are received and converted into electrical signals in the receiver. Due to the interaction of the ultrasonic oscillations with heterogeneities in the gas flow secondary oscillations becomes phase-modulated. The modulation signal is allocated by the phase detector. Based on this signal the time of movement heterogeneity between sections is determined. Calculator performs the same function as in the first method. The advantage of this method is the relative simplicity and reliability of the physical method, no effect the gas composition on the result of flow velocity. Disadvantage – a narrow dynamic range.

3. Field Studies

The results of field studies of several ultrasonic flowmeters carried out at various times and on different oil production objects are presented in this paper. Tests of pulse-time GiperFlow-US, GUVR-011A2, FlowSic 100 flowmeters and Doppler flowmeter Dymetic-1222 were carried out.

The main parameters of ultrasound flowmeters are shown in Table 1.

Table 1. Main Parameters of Ultrasound Flowmeters

Parameter	Value			
	GiperFlow-US	GUVR-011A2	FlowSic 100	Dymetic-1222
Dynamic Range	1:100	1:350	1:4000	1:50
Flow Velocity, m/s	0.3 to 30	0.1 to 35	0.03 to 120	
Volumetric Gas Discharge, m^3/h	5 to 200 000	1 to 10 000		16 to 30 000
Pipeline Diameter, mm	100 to 1 600	50 to 400	100 to 1 800	100 to 600
Intrinsic Relative Error, %	≤ 2.5	≤ 1.5	≤ 1	≤ 2

Figure 1 shows the described above flowmeters on various oil and gas production objects.

Figure 1. In order: GUVR-011A2, FlowSic 100 flowmeters and Dymetic-1222 flowmeter.

GiperFlow-US flowmeter was installed in a horizontal pipeline of feeding APG from the gas separator at oil production object.

During the measurement, analysis of the results, as well as inspection of the measuring area after dismantling was concluded that the oil is ejected periodically in the APG pipeline. It is connected with a lower location of the separation tank compared with the technological tank of the oil accumulation. These points of emission are marked by peaks of high flow rate at Figure 2.

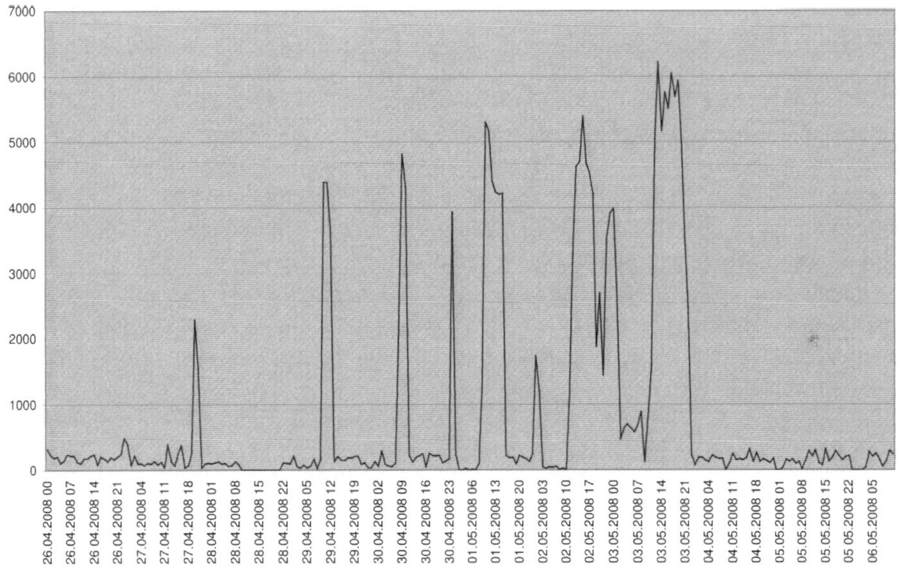

Figure 2. Flow rate volume measured by GiperFlow-US.

A significant excess of the measured flow rate versus the actual APG flow rate explained by the change of the medium (APG) to the denser medium (oil).

Following results of gas flow rate using an ultrasonic flowmeter HyperFlow-US were obtained: value of the hourly flow rate Q_C varied in the range from $Q_{MIN} = 1.5$ m^3/h to $Q_{MAX} = 400$ m^3/h (except for oil ejection in some measurement periods). Under these conditions, a more accurate measurement should be carried out using of measuring instruments are less sensitive to contamination of the APG.

Also, the measuring system based on ultrasonic flowmeter GUVR-011A2 was manufactured (Figure 1a). This measuring system was used to determine gas flow rate to the flare pipe. The results of measurement are shown at Figure 3a.

IOP Conf. Series: Materials Science and Engineering **86** (2015) 012020 doi:10.1088/1757-899X/86/1/012020

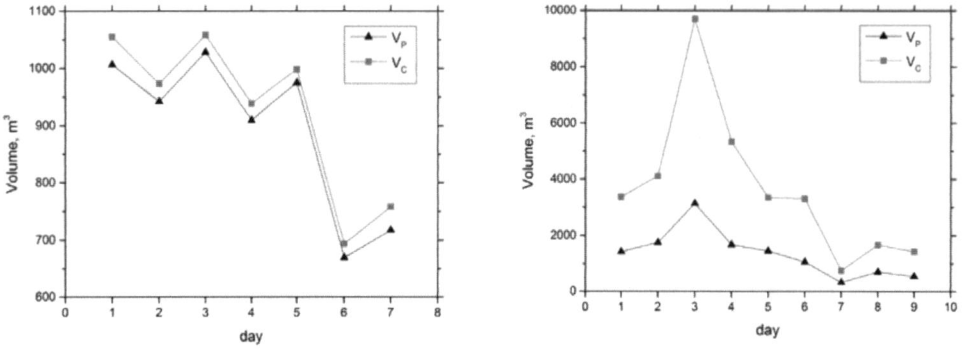

Figure 3. The results of measurement: a) gas flow rate at the flare pipe; b) at the pipeline of feeding APG from the separator, there V_P – volumetric gas discharge at the day, V_C – volumetric gas discharge reduced to standard conditions.

At this object the GUVR-011A2 flowmeter may be used. Value of APG flow rate represent the actual flow rate.

This measuring system was used later to determine APG flow rate on gas supply pipeline from the separator on another oil production object. At the Figure 3b the results of daily measurements and calculations for 9 days are presented. Due to the significant pollution acoustic sensors of the flowmeter data spread of measurements and calculations is observed. Therefore to use the GUVR-011A2 flowmeter on the given object should be excluded getting the condensed fraction of the APG in the measuring pipe.

The analysis of more than 20 measuring systems (at various oil production objects) based on the FlowSic 100 ultrasonic flowmeter was carried out.

The advantage in terms of organization constant account the APG – high reliability (including in the northern regions) and the accuracy of the determination the flow rate, minimizing costs for maintenance and routine maintenance. Disadvantage – the high cost and, accordingly, the high cost of repairs in case of failure.

Finally, we consider the ultrasonic flowmeter Dymetic-1222. Fitting the flowmeter and related pressure and temperature sensors made in a horizontal pipeline of the oil production object (Figure 1a). Measurements were made within 12 days from the launch date. The results are shown at Figure 4.

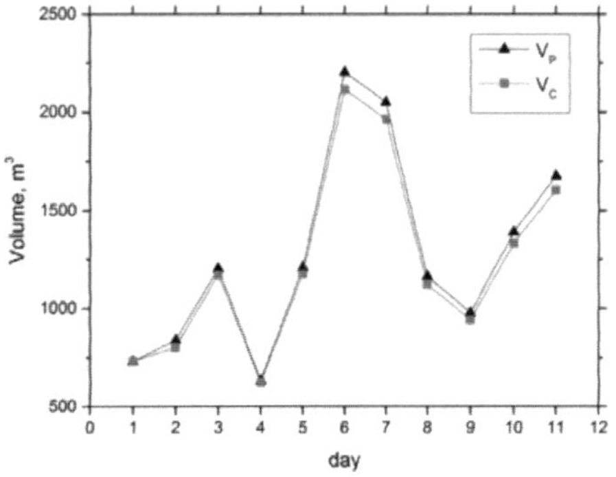

Figure 4. The results of measurement the gas flow rate obtained by Dymetic-1222, there V_P – volumetric gas discharge at the day, V_C – volumetric gas discharge reduced to standard conditions.

From the measurement results it can be concluded that they correspond to the actual flow rate values.

Advantages of Dymetic-1222 flowmeter – stable operation in polluted medium, a relatively low price. Disadvantages – big time lag when the actual flow rate is changing, weak mathematical apparatus processing the primary signal from sensors, and a relatively high percentage of failures due to flaws components and software.

5. Conclusions

The presence of contamination in the APG flow at temperatures above the dew point no creates of difficulties for most flowmeters (in terms of working algorithms). The exception is the presence in the APG flow of solid particles (e.g., sulfur), which can disturb the operation of sensitive elements some models of flowmeters and cause failure of the device.

But if the APG temperature drops below the dew point, the hydrocarbons begin to condense in the flow (until the appearance the liquid phase) and at the sensors.

At still lower temperatures, liquid fractions crystallize and deposit in the form of a solid crust on the pipeline and sensors. In addition distorting the measurement results, this leads to the failure of some models of flowmeters. In exceptional cases, this leads to a complete APG flow shutoff.

Also aggressive medium of APG may negative impact on the sensors. If the sensor is not protected metallized flask, corrosion and destruction the sensitive part of sensor can occur. In addition, the presence of a second phase (condensed heavy hydrocarbons or emergency ejected oil and other oil fractions) distorts received primary signal. Therefore, the accuracy of accounting in these conditions becomes questionable. Figure 5 shows the flowmeter sensors are polluted with oil.

Figure 5a. Polluted sensor of FlowSic 100 flowmeter.

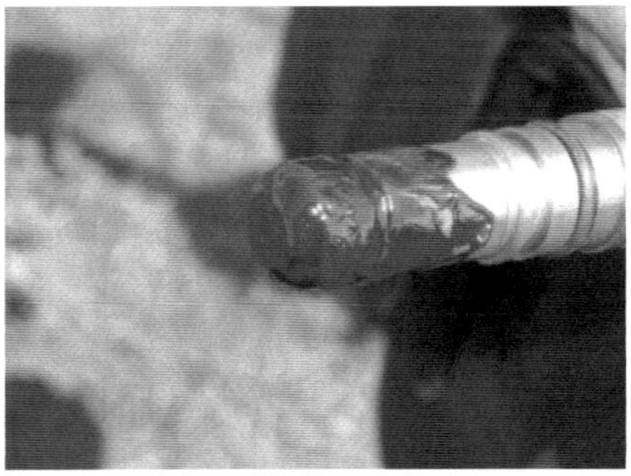

Figure 5b. Polluted sensor of GiperFlow-US flowmeter.

Approach to solving these problems an individual. Somewhere solution is using of gob deflector, somewhere installation of additional separator, somewhere changing technological line, or transfer the flowmeter to another location.

In conclusion, it should be noted that the use of ultrasonic flowmeters is limited to the high cost. It is starting from 30 000 euros for set.

The work is performed according to the Russian Government Program of Competitive Growth of Kazan Federal University.

References

[1] GOST R 8.733-2011 *System parameters and measurements of the amount of free gas. General metrological and technical requirements* (Moscow: Standartinform) p 32

[2] GOST R 8.615-2005 *Measuring the amount extracted from subsurface oil and gas. General metrological and technical requirements* (Moscow: Standartinform) p 23

[3] Fazlyyyakhmatov M G, Kashapov N F, Khayritonov Kh A, Lazarev D K and Lazarev V K 2014 *IOP: Conf. Ser. Material Science and Engineering* **69** (1) 012033

[4] Sagdiev R K, Denisov E S, Evdokimov Yu K, Fazlyyyakhmatov M G and Kashapov N F 2014 *IOP: Conf. Ser. Material Science and Engineering* **69** (1) 012012

[5] Denisov E S, Temyanov B K, Sagdiev R K and Fazlyyyakhmatov M G 2014 *IOP: Conf. Ser. Material Science and Engineering* **69** (1) 012014

IOP Conf. Series: Materials Science and Engineering **86** (2015) 012021 doi:10.1088/1757-899X/86/1/012021

Hardening of the surface plasma jet high-frequency induction discharge of low pressure

N F Kashapov and S N Sharifullin
Kazan Federal University, 420008, Kremlyovskaya str., Kazan, Russia

E-mail: Saidchist@mail.ru

Abstract. The work presents results of research on the hardening surfaces of the products and increase their roughness class of plasma jet of high frequency induction discharge of low pressure. It is shown that such processing allows to clear at the same time a surface of all types of pollution, to remove a defective layer after its machining, to receive a uniform microstructure, to raise a roughness class on 2 – 3 units.

1. Introduction

There is art, where hardness and surface roughness class products are fundamental determinants of its technical characteristics. For example, in shipbuilding and aircraft such characteristics are the weight of the vessel and its aerodynamics. From them, depend primarily on the tactical and technical properties of this type of machine. Available conventional processing techniques do not allow to obtain higher surface roughness grade 12 and thus not increase its hardness. Therefore, as of today, they can not meet the challenge of competitive products on the world market. There need innovative approaches and use of unusual and breakthrough technologies. These technologies can be only high technology that can produce high-quality products, promising new materials, coatings and surfaces with unusual properties. These include plasma gas discharge technology.

2. Materials and Methods of Research

The technique of gas discharge is used in atmospheric and low pressures. Low-pressure discharge differs from the discharge of atmospheric pressure spike of the average kinetic energy of charged particles in the direction of increasing, especially energy electrons. By blowing a gas discharge plasma gas obtain plasma jet, the average temperature of which may be only a few hundred degrees Celsius, while the electrons – a few tens of thousands. Depending on the method of preparation of the discharge plasma are distinguished as the plasma arc, spark, corona, glow, microwave and RF discharge. Respectively called and apparatus for producing discharges. Plasma gas can be any gas, as inert and aggressive.

A wide range of energy, thermal and gas-dynamic characteristics of low-temperature gas-discharge plasma allows its use in a variety of technologies [1-4]. In this case, the thermal power of the plasma jet can be varied from tens of watts to tens of megawatts, the temperature – from several hundred degrees to 50 000 °C, the effective efficiency Heating may be up to 90%, of the plasma jet velocity - 1000 m/s, continuous operation – up to 1000 hours [5].

The surface microstructure of samples before and after treatment HFI-flow low-pressure plasma was investigated in an electron microscope UEMV-100A. The microhardness of the surface is determined by the PMT-3. Measurements of light scattering glass samples were performed on non-standard measurement of the scattering laser system LUIR based on a comparison of scattering coefficients of samples and the reference plate. Radiation resistance surface samples was determined on a special stand with a laser LTI-5.

3. Results and discussion

In our case, the choice of the form of the plasma to perform this task, you must meet the following requirements:

1. The plasma must be gas-discharge;
2. The plasma must be clean, ie. E. Products without erosion of the electrodes;
3. Under the influence of temperature of the plasma should not change the basic physical and Mechanical properties of the material of the workpiece;
4. to a plasma properties of the surface layer of the items should significantly change in the direction of improving the quality;
5. To remove the products of processing to use the plasma jet.

Analysis of the plasma properties and possibilities of its application is shown that these requirements can be satisfied high-frequency induction discharge (HFI), low-pressure inert gas stream Fig. 1 [6–20].

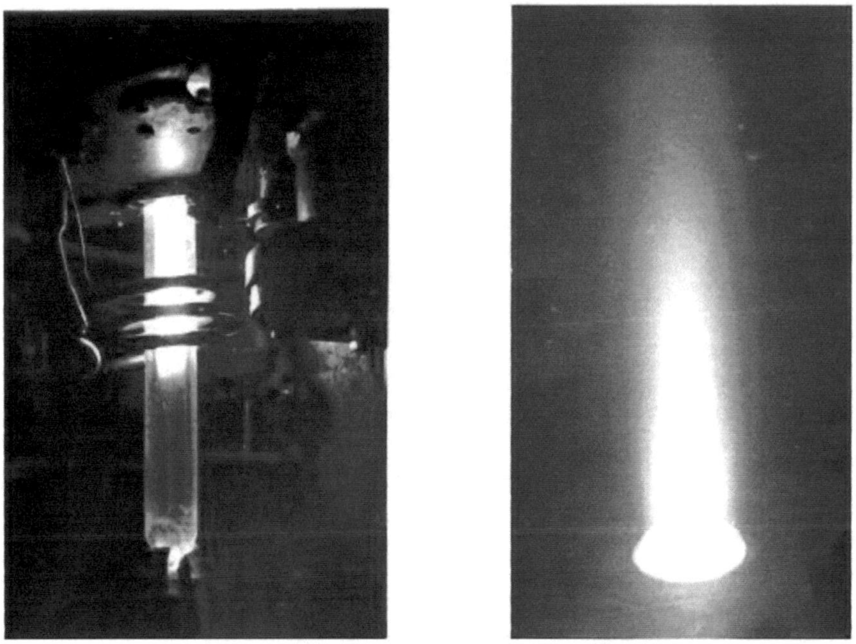

Figure 1. Photographs HFI-discharge (a) and the plasma jet (B) at low pressure

Studies have allowed to define treatment regimens HFI-low pressure plasma for the task. Were treated surfaces of materials of aluminum, titanium, steel, copper and glass of different brands. Below we present some results of the research. Fig. 2 shows photographs of the surface microstructure K8 glass before and after treatment with the HFI-flow low-pressure plasma produced by an electron microscope UEMV-100A. Comparison of the photographs shows that the existing micro-defects surface after plasma treatment are disappearing and formed a brand new, smoother, more uniform surface microstructure. The analysis of these images and the results of other studies led to the conclusion that during processing high frequency induction plasma samples of low pressure is removed the defective surface layer and its subsequent polishing. We point out that the surface roughness of the samples to the plasma treatment is equal to 10 – 11-th class. After treatment reaches

13 – 14 grade surface roughness. Note that the table class roughness on GOST finishes 14th class. It is theoretically achievable roughness class. In practice, existing processing methods do not enable the achievement of higher roughness class 12th.

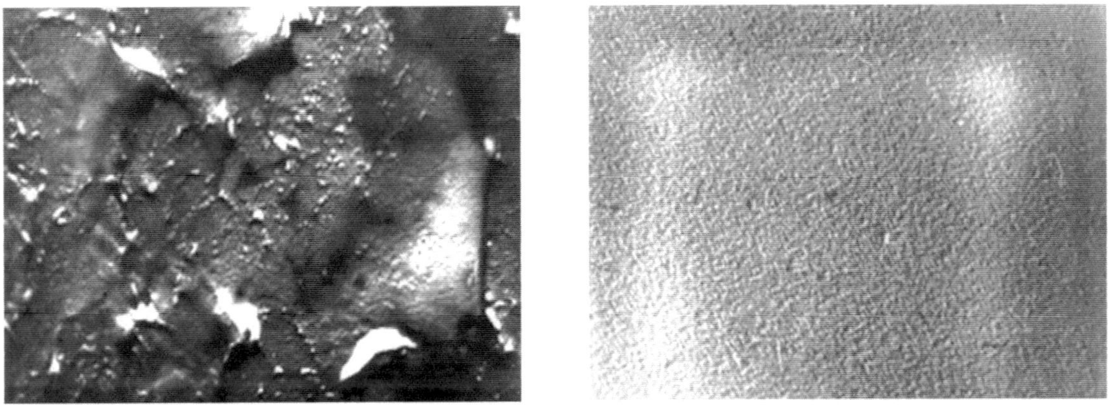

a б

Figure 2. Photos surface microstructures K8 glass before (left lane) and after (right lane) processing flow HFI-low-pressure plasma (x20 000 in the format of A4 paper, photo width corresponds to 4 micron of the surface)

Carrying out the surface treatment of the samples of aluminum, titanium, steel and copper, obtained similar results as that for glass materials. Consequently, no matter what material is manufactured item. The effect is the same. If only the material parts withstand temperatures of 100 - 200 °C. As an example, Fig. 3 shows photographs of a microstructure of the sample surface of the titanium alloy VT6 before and after treatment with HFI low pressure plasma. They show that after plasma treatment on the surface of the material is practically no traces of machining, and dramatically increased its class roughness.

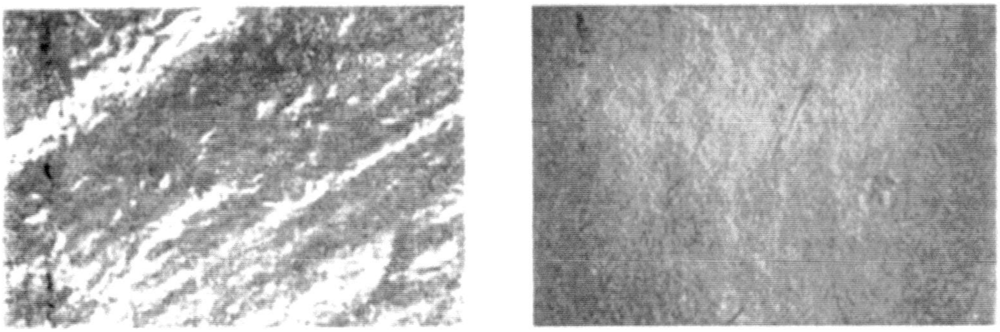

Before After

Figure 3. Photographs of microstructures of the titanium alloy BT6 before and after treatment with the HFI-flow low-pressure plasma (x20 000 in the format of an A4 page)

Studies have shown that exposure to a plasma of metals other than cleaning and polishing the surface, and can significantly alter the properties of the surface layer. Measurements of the PMT-3 showed that the HFI-plasma treatment under low pressure leads to an increase in the microhardness of the surface.

For example, the sample of aluminum alloy D16T it rose from 520 to 1140 MPa, steel Article 45 – from 2500 to 5000 MPa and a titanium alloy VT6 – from 2600 to 4600 MPa. Interesting is the fact that under normal conditions of storage within two months of the above material D16T led to an increase in its microhardness up to 1500 MPa. One must assume that under the influence of the plasma flow structural changes occur in the surface layer of the metal, the ordering of structural cells, burn-in and as a result its hardening. The result is a modified surface layer having high hardness, homogeneous microstructure and high grade roughness.

With respect to the defective surface layer. Whenever the machining of pieces in the surface layer of introduced kinds of pollution dust, adsorbed gases, water vapor, oil, particulates chips, abrasive particles, polishing paste, etc.. Pr. Same dirt remain in the surface roughness after machining. It is known that as a result of machining the surface of a so-called defect layer [8]. It is a surface layer consisting of a rough layer and microcracks. The depth of the defect layer during rough machining may be up to 50 microns. Roughened layer is about a quarter of the total thickness of the defect layer. Complete removal of the defect layer surface of the part of any, even the fine machining, is impossible, since she will leave behind a defect layer.

In order to ensure the removal of the defect layer and the surface roughness decrease its exposure to the plasma, were investigated by measuring the light scattering coefficient and the radiation resistance of optical glass surfaces before and after treatment with the HFI-flow low-pressure plasma. By passing a light beam through the glass core energy loss depends on the dirt and the roughness of its surface. Therefore, these numerical parameters will specifically describe the effect of plasma on the surface of the material.

Light-scattering measurements were carried out on a non-standard installation of laser scattering measurement LUIR based on a comparison of the scattering coefficients of the samples and the reference plate. Scattering was determined at a wavelength of 632.8 nm in the «S» - polarized laser radiation. Index «S» means the polarization when the electric vector is perpendicular to the plane of incidence. The results of measurements of the scattering coefficient for the 14 glasses K8 showed that HFI-plasma treatment at low pressures allows reducing 2-3 times the scattering coefficient of glass.

Radiation strength surface samples was determined on a special stand with a laser LTI-5. Measurements have shown that after the HFI-plasma treatment the radiation resistance of the surfaces of glass K8 increased by 1.5 – 2 times. The decrease in light scattering coefficient of optical glasses is 2 – 3 times after the HFI-plasma treatment of their surfaces, with a simultaneous increase in the radiation resistance of 1.5 – 2, is convincing proof removal of defective surface layer and increase its roughness class.

JavaScript removal of defective surface layer HFI plasma treatment can be used to increase the transparency of glass cockpit fighter with a simultaneous decrease in the angle of inclination. This in turn will improve the aerodynamic characteristics of the aircraft.

By varying the length of the surface treatment, it is possible to explore the process of changing the microstructure. Fig. 4 shows photographs of the microstructures of the sapphire surface at different durations of treatment. Comparison of the photographs shows that plasma

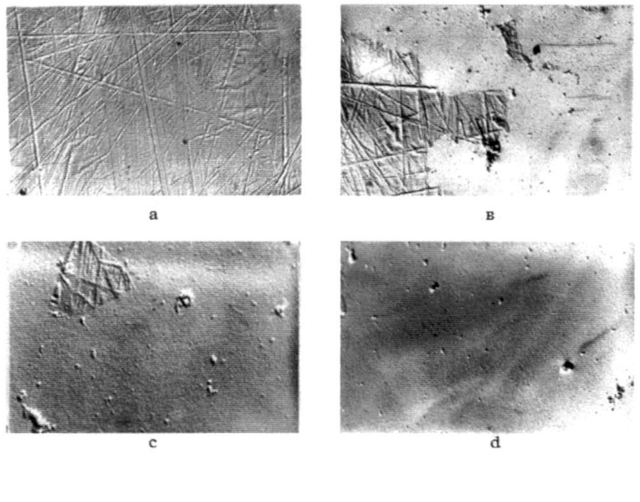

Figure 4. Changes in the microstructure of the sapphire surface, depending on the duration of treatment HFI-plasma (x20 000 in the format of an A4 page)

treatment at first there is the fragmentation of large micro defects into smaller and then removing them from the surface of the plasma stream. Processing occurs due to bombardment of the surface by ions, electrons, excited atoms, photons, and as a result of mechanical and thermal effects of the HFI-plasma flow, ie surface of the sample is under the influence of a complex energy, thermal and the gas-dynamic parameters of the plasma flow. Studies have shown that the surface treatment of the HFI-plasma subject to any details.

4. Conclusions

Analysis of the results for the HFI-plasma surface treatment of parts made of different materials allows us to conclude that there is a new tool for modifying a thin surface layer of the part. It can simultaneously clean the surface of all types of contaminants to remove a defective layer after the machining to obtain a uniform microstructure, and to increase the microhardness of 1.5 – 2 times.

References

[1] Galyautdinov R T, Kasparov N F and Luchkin G S 2002 *Inzhenerno-Fizicheskii Zhurnal* **75** 170–173
[2] Fadeev S A, Kashapov N F and Larionov V M 2014 *IOP Conf. Ser.: Mater. Sci. Eng.* **69** 012006
[3] Fadeev S A and Kashapov N F 2013 *J. Phys.: Conf. Series* **479** 012009
[4] Kashapov N F, Saifutdinov A I and Fadeev S A 2014 *J. Phys.: Conf. Series* **567** 012004
[5] Dautov G Y, Dziuba V L and Karp I N 1984 *Plasma torches with stabilized arcs* (Kiev, Naukova Dumka) 168 p
[6] Sharifullin S N 1977 *Research and application of some high-frequency low-pressure discharge. (Diss. cand. tehn. Sciences)* (Kazan) 198 p
[7] Kashapov N F *2001 Inkjet RF plasma torches during the coating process under dynamic vacuum. (Diss. Doctor. tehn. Sciences)* (Kazan) 506 p
[8] Israfilov I H, Samigullin A A, Sharifullin S N and etc. 1976 *The energy characteristics of high-frequency plasma unit at low pressure (Physics and Chemistry of Materials Processing, # 3)* Pp 26-31
[9] Gainutdinov I S, Dautov G Y, Samigullin A A, Sharifullin S N, Shcherbakov V D 1977 *Cleaning and polishing of substrate surfaces high frequency induction plasma of low pressure* (*Physics and Chemistry of Materials Processing, # 6*) Pp 150-152
[10] Abdullin I Sh, Kashapov N F and Kudinov V V 2000 *Changes in the structure and composition of the surface of steel and titanium alloys under the influence of high-frequency low-pressure discharge* (*International scientific journal "Advanced Materials,# 1)Pp* 56-60
[11] Abdullin I Sh and Kashapov N F 2000 *Modification of steels and titanium alloys by means of high-frequency low pressure plasma (Preprint, KSTU)* 16 p
[12] Abdullin I Sh and Kashapov N F 2000 *Preparation of surfaces of metals, dielectrics and semiconductors to the coating by means of non-equilibrium low pressure RF plasma (Preprint, KSTU)* 16 p
[13] Abdullin I Sh, Kashapov N F and Kudinov V V 2000 *Treatment of inorganic materials nonequilibrium low-temperature plasma before coating (International scientific journal "Advanced Materials", # 3)* Pp 88-94
[14] Kashapov N F 2000 *Changes in the structure and composition of the surface of steels and titanium alloys after exposure to low temperature plasma reduced pressure (Hands. Depon. in ONTI union VINITI, # 973)* (V00, KRTU, Kazan)
[15] Denisov E S, Temyanov B K, Sagdiev R K and Fazlyyyakhmatov M G 2014 *IOP Conf. Ser.: Material Science and Engineering* **69** (1) 012014

[16] Abdullin I Sh, Sharifullin S N, Shcherbakov V D and Vdovin D A 1979 *Treatment of metal surfaces induction plasma flow* (*Low-temperature plasma: Intercollege. Sat.)* (Kazan: KAI) Pp 55-58

[17] Abdullin I Sh, Sharifullin S N et al. 1981 *Study changes in the surface properties of the titanium alloy under the influence of the induction plasma diffuse discharge* (*Low-temperature plasma: Intercollege. Sat.)* (Kazan: KAI) Pp 18-22

[18] Sagdiev R K, Denisov E S, Evdokimov Yu K, Fazlyyyakhmatov M G and Kashapov N F 2014 *IOP Conf. Ser.: Material Science and Engineering* **69** (1) 012012

[19] Sharifullin S N 2014 *Hardening of surfaces of products and increasing their roughness class of low-temperature plasma jet (Abstracts of the II International scientific and practical conference "Perspective materials and technologies for production" in the framework of the International Forum "Crimea Hi-Tech - 2014")* (Sevastopol, 22-28 September 2014. M .: NUST "MISA") Pp 282-288

[20] Sharifullin S N 2009 *Improved operational reliability high pressure fuel pumps automotive diesel engines (Diss. Doctor. tehn. Sciences)* (Moscow) 368 p

IOP Conf. Series: Materials Science and Engineering **86** (2015) 012022 doi:10.1088/1757-899X/86/1/012022

Triboplasma - the main component of plasma-chemical processes occurring in the friction units of machines and mechanisms in the presence of these drugs Tribo

N F Kashapov and S N Sharifullin
Kazan Federal University,
42000818, Kremlyovskaya str., Kazan, Russia

E-mail: Saidchist@mail.ru

Abstract. In this paper, a technology capable of repairing worn surfaces of the friction units of machines and mechanisms. This technology can successfully be used in any production of engineering products. The proposed technology is low-cost, efficient and easy to use. It is shown that the processes occurring in the friction units of machines and mechanisms in the presence of these drugs tribo triboplasma plays a fundamental role, which arises in the contact mating surfaces.

1. Introduction

At present, the state technical policy aimed at the modernization of worn equipment, including agriculture, through the application of high-performance technologies, called nanotechnology. However, the power of this modernization, says a lot, but in reality little has been done. Under the modernization of equipment worn understood its restoration to the achievement of the above parameters passport.

In this article we will focus on the technology dimension to rebuild worn surfaces of friction units of machines and mechanisms. The foundations of this technology has been incorporated in the 80-ies of the last century. It was used in the serpentinite Tribo drugs basis. These were named RVS Tribo drugs additives and surface treatment technology of friction units with their application - RVS technology. The essence of this technology is that the RVS additive is introduced into the standard oil or grease. The preparation was then transported to the oil on the surface of the friction units and under the influence of pressure, temperature and triboplasmy create conditions for the return of wear on the worn surface of the seat with a partial or full recovery of the geometric dimensions of parts [1–4]. When this occurs the formation of layers with a high wear resistance and low friction. The result is a so-called no-wear operation of equipment and machinery actually become later a new passport with the parameters of the above values.

This technology was based on the so-called opening of "The effect of low friction surfaces of friction units in the presence of layered silicates." The availability of this technology, simplicity, low cost, and the economic benefit it allowed widespread development of various kinds of practices experimentalists. But it is almost to the present day has not been studied mechanism of wear-resistant coatings on the surfaces of friction units and aggregates in the presence of Tribo drugs not conducted in-depth research on the study of processes occurring at the same time. So do not otrabotanom PBC technology, its use of "blind", sometimes led to negative results. Hence, some caution in the technology and the lack of its large-scale application.

Should not be confused with each other, "RVS additives" and "Additives" [1]. Unlike addition of additives commonly known that when the additive oil change requiring repeated application, and additives remain on the friction surface and continue to work when replacing the oil. Experience has shown that in this case the effect of the treatment, depending on the operating conditions, for example for cars continues to operate from 50 to 100 thousand miles of their run.

There are several versions of the mechanism of forming a coating on the surfaces of friction units at their processing RVS additives. On one of them, supplements, getting oil on the friction pair, activate the defective surface. Under the action of catalysts and triboplasmy RVS additive cleans the friction surfaces of the layers, oxide films and fragile structures deformed crystals. At the same time on the cleaned surfaces appear free molecular bonds, which as "octopus" attract crystals additives and wear products. Begins the formation of new crystals in the valleys of roughness, accrued on the crystal lattice of the surface layer of the metal. In the future, these crystals are oriented along the field and coalesce to form over the entire surface of the contact point diamond-like coating. I would like to draw attention to the fact that with all this catalysts RVS additives remain in the gaps of friction units and how - to "duty." In the event of further deterioration surface they again begin the process of recovery. Thus, there is no exploitation of the wear art.

It is necessary to say more about the following. The surface roughness is characterized by its peaks and valleys. Depressions on the surface can be represented as an area of wear. Friction modifier, getting into the zone is captured and held by the metal atoms having free communication. During the loading energy activator directs excess energy to create a new crystal lattice. After a while the depression is filled with diamond-like coating. As soon as the surface of the cavity due to the coating compared with the projection, the energy conditions necessary for recovery. So the item, will never gain in the amount of more than the seizure will not happen mating surfaces. The system has stabilized.

2. Materials and Methods of Research

In today's world there are many different additives, including the Russian production. Raw material for the manufacture of additives is essentially opaque mineral called serpentinite. Before it becomes a resursoenergosberegayuschuyu additive initially undergoes a special milled to nano-sized, and then filtered of impurities. The basis of the serpentinite is serpentine. Available today for the 18 varieties of serpentine, in practice, mainly used antigorite, chrysotile, lizardite-1T. They are described by general structural formula $Mg_6 [Si_4O_{10}] (OH)_8$, but are identified by x-ray diffraction is different. For example, in the analysis of drugs in PBC diffractometer XRD 6000 library spectra of the International Institute of ICSD shows them to 12 slightly different minerals. [1, 2] Component composition of serpentine: MgO - 43%, SiO_2 - 44% H_2O - 12.9%. In the serpentine also contains about 13% of the constitutional water; constitutional water is represented in minerals hydroxyl ions (OH), and in rare cases with H + ions, ranging in lattice sites. This mineral water is held firmly at room temperature, but is released during heating in the temperature range of 300 - 1300 °C. Allocation of water is accompanied by the destruction of the crystal lattice of the mineral with an additional release of large amounts of thermal energy. However, in the serpentinite rock includes several varieties serpentine, magnetite inclusions, chromite and accompanying plurality of minerals based on iron, aluminum, calcium, nickel, and some other elements. In serpentinites different fields all this "walking" in a large range. Therefore, a more general formula serpentine Tribo drugs, taking into account their variable composition is: $(MgO)\ x\ (SiO_2)\ y\ (H_2O)\ z$, where x, y, z - variables.

Was held X-ray fluorescence analysis of the elemental composition of eight types of powders and supplements RVS source of serpentinite [5]. The analysis showed that in all the major chemical additives RVS elements are magnesium, silicon and iron. In almost all formulations the presence of aluminum, calcium, chromium, manganese and nickel. RVS supplements from different manufacturers have about the same elemental composition. By quantitative content of these additive elements, taking

IOP Conf. Series: Materials Science and Engineering **86** (2015) 012022 doi:10.1088/1757-899X/86/1/012022

into account the errors have a close. In some compositions, as well as on data from Nanocentre GOSNITI in small amounts include sulfur, copper and zinc. Apparently, the producers included in the data elements based on their research results. It should be noted that the addition of LLC «Tehnodar" (Karelia) is very different from the elemental and quantitative composition of all the additives that may be due to the difference of the feedstock. At the same serpentine are almost all of the above elements.

3. The mechanism of action of repair and restorative composition in the friction units of machines and mechanisms

In recent decades, there have been some studies that reveal the essence of the processes taking place in the friction units of machines and mechanisms in the presence of additives in them RVS. It turned out that these processes play a major role in the emerging contact points mating surfaces friction unit electrical discharges. In recent decades, there have been some studies that reveal the essence of the processes taking place in the friction units of machines and mechanisms in the presence of additives in them RVS. It turned out that these processes play a major role in the emerging contact points mating surfaces friction unit electrical discharges. The question arises. From these ranks, and what role they are given in the PBC technology?

Sparks impact and friction has long known. Only a few people delved into its nature. In such technique called spark friction sparks. They are the result of the transition of the mechanical energy into heat. With relative sliding of the two bodies due to deformation of the material involved in the formation and destruction of the friction relationships in the thin surface layer formed by heat. Increased temperature can thus cause local softening and melting of the material. The heat spreads from contact spots deep both contacting bodies, the heat flows are allocated depending on the thermal properties of contacting bodies, the size and conditions of the heat sink. [6] At steady-state thermal regime mechanical energy transformed in micro surface layer into heat is transferred into the interior of the mass of the rubbing bodies by conduction is not completely. Part of it is a result of structural transformations in the surface layers, and thermionic emission, thermal diffusion, and so on. Dissipated directly to the environment [6]. When a rough surface slides on the other, the contact occurs only at certain points which are quite a short time in contact. The lifetime of a single spot can be $10^{-7} – 10^{-8}$ s [7]. On a single spot having short-term individual temperature flashes. Flash can reach considerable size, which affects the conditions of interaction of the rubbing surfaces. As a result, a very short period of time, a thin layer of molten material, and even in the high-energy state, similar plasma or magma. This "MAGMA-plasma" later called "Triboplasma."

According to the model "Triboplasmy" "released by mechanical action on the material can greatly exceed the energy heat melting due to poor thermal conductivity of solids not only leads to the melting of the substance but also to its sublimation, and then rise to a state in which it is in the form of ions and electrons (plasma state). For such applications, the laws of classical thermodynamics are not met; a thin surface layer of material is converted, resulting in a collision zone formed irregularities plasma (triboplasma); the process is accompanied by the emission of electrons.

The approach to the description of features of the structure and behavior of matter tribomating in a state can be considered triboplasma idea of anomalously enhances the process of diffusion mass transfer in friction associated with changes covering the structure of the surface layers of friction units [8]. At the same time observed as transformation has been accompanied by structural defects, and the reverse processes. On fault detection surface phenomenon special mark superimposed thermal processes, in particular, the diffusion coefficient D_0 corresponding triboplasme can be calculated using the relation:

$$D_0 = \frac{R_Д^{\;2}}{h}(Q^* - W_o)\text{; } [D_0] = \left[\frac{m^2}{s}\right], \quad (1)$$

Where $R_Д$ - Debye radius, m; h - Planck constant, J/s; Q* = kT + Q - quantity characterizing the kinetic energy of the thermal motion of particles triboplamy, J.; k - Boltzmann constant, J / K; T -

Temperature triboplasmy, K; Q - the heat released by friction, J; W_∂ - activation energy of diffusion processes, J.

From this relation it follows that the thermal activation surface greatly increases the activation energy of diffusion processes and leads to abnormal strengthening of the diffusion coefficient. In further calculations we can obtain an expression that reflects the proportionality of the diffusion coefficient of the square root of the temperature field of the frictional contact. From this it can be assumed that triboplazma with properties close to gases. Therefore, it can be described by an equation similar to the low-temperature gas-discharge plasma [9-16].

4. Results and discussion

In general, laboratory and production testing pioneers RVS-technology form the basis of a fundamentally new direction in tribotechnology - restoration of the worn surfaces of interfaces to their nominal values, and in some cases above, under the influence of natural mineral additives introduced in tribosredu.

In 2012, our high school together with "RealInProekt" (St. Petersburg) has developed its own PBC supplement on the basis of serpentinite, entitled "TATGEO" to upgrade the worn equipment. Appointment of friction modifier "TATGEO" - restoration of worn surfaces of friction units of machines and mechanisms without disassembly with a further resource of their work up to 2 thousand. Moto hours or up to 100 thousand. Kilometers car. Our friction modifier from existing analogues differ by more than ten items, including feedstock composition, lower cost manufacturing technology and application.

A couple of words about the source of raw materials and manufacturing techniques.

Manufacturers PBC additives feedstock collected ore deposits. However, not everyone is suitable for PBC serpentine powder. Distinguish mikrocheshuychaty serpentinite sheet called antigorite, fine fiber, called chrysotile, and fine-grained, called lizardite. Based on many years of experience our experts choose the most effective serpentinite. It is with the addition of chrysotile antigorite. After grinding us RVS powder added in certain proportions various surfactants, inhibitors, scavengers, stabilizers, and catalysts.

The scope of surface treatment technology of friction units tribo drugs:

- The machinery industries (rolling and sliding bearings, shafts, bearings and Screws, etc.);
- Motor vehicles (engines of all types and sizes, gears, joints, the hinge of equal angular speeds, mechanical gearbox);
- Rail transport (engines, gearboxes, wheel-rail pair, etc.);
- Sea and river vessels (engines, diesel generators, etc.);
- Weapons and military equipment.

Achieved the effect of surface treatment of friction units tribo drugs:

- Increased service life of parts and machinery in 2 - 3 times;
- Increasing the power of internal combustion engines by 10 - 15%;
- Reduce energy consumption by 10 - 20% (machine tools, generators, blowers);
- Reduced consumption of fuel and lubricants for 7 - 15% (internal combustion engines, bearings, gears);
- Increased oil change intervals of oils and lubricants 1.5 - 2.5 times;
- Reduction of CO and C x H y in the exhaust gases of internal combustion engines in the 2 - 3 times;
- Improving the performance characteristics of military equipment, including small arms;
- Reduce the cost of production of parts, repair and maintenance costs in the 1.5 - 2.0 times.

5. Conclusions

1. The electrical discharge that occurs in friction machinery and equipment refers to the section of low-temperature plasma.
2. Triboplasma equations can be described as applied to low-temperature plasma.

3. Research triboplasmy is relevant not only in terms of scientific interest, but also in terms of practical significance.

4. RVS is designed to supplement the basis of serpentinite, entitled "TATGEO" for modernization of worn equipment. It differs from existing analogues for more than ten items, among which the feedstock composition, lower cost manufacturing techniques and applications.

References

[1] Dunaev A.V. Modernization of equipment worn with tribopreparatov. Monograph /A.V. Dunayev, S.N. Sharifullin. - Kazan: Kazan. University Press, 2013. – 272 p.

[2] Anatoly Dounaev. Friction Surfaces Modification Using Tribo-Compounds /Anatoly Dounaev and Said Sharifullin //World Applied Sciences Journal. ISSN 1818-4952, 2014. – Vol. 31 (2). – Pp. 272 – 276.

[3] Sharifullin S and Rogozhkin N 2014 *IOP Conf. Ser.: Mater. Sci. Eng.* **69** 012039

[4] SN Sharifullin Modernization of worn nodes and mechanisms of friction with nanotechnology /S.N. Sharifullin // Abstracts of the II International scientific and practical conference "Perspective materials and technologies for production" in the framework of the International Forum "Crimea Hi-Tech - 2014", Sevastopol, 22 - 28 September 2014 - M .: NRTU "MISA" , 2014. - S. 277 - 282.

[5] Sharifullin S.N. Some studies of the repair and recovery formulations /N.R. Adigamov, S.N. Sharifullin, I.H. Rizvanov, V.M. Babaev //Works GOSNITI, 2013. – Volume 112. – Part 1. – Pp. 26 – 33.

[6] KRAGELSKY I V 1968 *Friction and wear. Ed. 2nd Revised. and add. (Moscow: Mechanical Engineering)* 480 p

[7] Garkunov D N 1985 *Tribotechnology (Moscow: Mechanical Engineering))* 424 p

[8] Dolgopolov K N 2009 *Improving the performance of household appliances through intensification processes lubrication of friction units (Author. Dis. for the degree of PhD. tehn. Sciences) (SRSUES, Mines)* 28 p

[9] Kashapov N F and Israfilov Z K 1991 *Journal of Engineering Physics* **60** 364–368

[10] Dautov G Y, Kashapov N F, Fayrushin I I and Egorova E A 2014 *Characteristics of the potential well for a non-degenerate electron gas in equilibrium dusty plasma (Russian Physics Journal. 4. Vol.57. # 3)* Pp 94-97

[11] Luchkin A G, Kashapov N F and Luchkin G S 2013 *J. Phys.: Conf. Series* **479** 012019

[12] Dautov G U, Kashapov N F, Fairushin I I and Dautov I G 2013 *J. Phys.: Conf. Series* **479** 012014

[13] Fadeev S A and Kashapov N F 2013 *J. Phys.: Conf. Series* **479** 012009

[14] Fadeev S A, Kashapov N F and Larionov V M 2014 *IOP Conf. Ser.: Mater. Sci. Eng.* **69** 012006

[15] Kashapov N F, Saifutdinov A I and Fadeev S A 2014 *J. Phys.: Conf. Series* **567** 012004

[16] Dautov I G, Dautov G U, Kashapov N F and Fairushin I I 2013 *J. Phys.: Conf. Series* **479** 012001

IOP Conf. Series: Materials Science and Engineering **86** (2015) 012023 doi:10.1088/1757-899X/86/1/012023

The use of SHS-process slag for the preparation of foundry sand

N Safronov, L Kharisov

Kazan Federal University, Naberezhnye Chelny Institute 423810, Naberezhnye Chelny, Mira av., 68/19, Russia.

E-mail: ln271@mail.ru

Abstract. The article presents the construction of a linear approximation of the response function (strength of molding sand on tensile strength and air permeability) in a given area by changing the investigated factors for which charge makeup of SHS-cast iron obtaining process and content of the slag product of this process in the molding mixture were used. The analysis of regression coefficients significance of the objective function was done to identify the most important input parameters.

1. Introduction

Plastic and liquid sand on cement binders have high technological properties, due to which they found quite widely usage in manufacturing steel and iron castings with particularly large mass [1]. Various structures of discussed molding compounds have been developed [1-5], which provide them with sufficient survivability and mobility, required strength and gas permeability, low crumbling, good knocking out ability, the optimal duration of curing. An integral and important component of the cement forming mixtures is the binder, for which special construction cement with high-coy specific surface area and the addition of gypsum are used. Construction cement from Pikalev plant and Vorovskoj plant is well proven [2]. Aluminous cement is widely used as a binding agent.

2. Basic part

In the present study the properties of the cement moulding mixture were investigated, in which process slag of obtaining SHS-cast iron was used as a binder [6,7]. This slag is similar to the aluminous cement and has binding properties as aluminous cement of grade 400, 500. The cost of used binding material is lower than of standard analogue, which is confirmed by other researchers [8], who also used the SHS process and received a floating fused alloy spinel.

Cement sand was made of the following components: quartz sand, slag process of obtaining SHS-cast iron, water, foaming agent (DS-RAS) in mass ratio 100:(13-16):8: 0,1. In the mixture there are no accelerators from adoption. This is due to the fact that the studied cement molding mixture has sufficient speed of binding material bondability [9]. The quality of the moulding mixture was evaluated on the strength parameters of its tensile strength and gas permeability.

To determine the strength of the molding mixture on tensile parameter samples in the core box were made by filling it with a mixture of easy seal by its hand. The excess mixture was cut off. The cured samples were assayed within 24 hours in normal conditions. The samples had the form of figure “8” with height 75 mm, thickness 25 mm, width of the head and neck 40 mm and 25 mm accordingly. Then the samples were subjected to stretching at a car company "Dietert" type 405.

To determine the gas permeability of sand, its right hand weighing 165±15 g was taken and placed in a one-piece steel polished liner with an inner diameter of 50±0.025 mm and height of 120 mm. In the appropriate seals samples were received with a height of 50±0.5 mm, which were dried in natural conditions within 24 hours. Next, the liner sample was mounted on the test apparatus for gas permeability. For this 2000 ml of air was poured through the sample under pressure 980,7 PA (10 cm water.ct.) and fixed the value of gas permeability on the readings. A machine Permmeter of "Dietert" firm was used as a setting for defining gas permeability of mixture.

The aim of the study was to determine a linear approximation of the response function (y1 – sand stretching strength, 105 PA; y2 – gas permeability, unit) in a given area by changing the investigated factors for which charge makeup of SHS-cast iron obtaining process (x1, x2, x3) and content of the slag product of this process in the molding mixture were used.

The charge of this process consisted of forge scale, siluminum and graphite shavings, powdered lime and ammonium nitrate. The composition of the SHS process charge – input controllable factors was formed in relation to the forge scale of the Foundry plant of OJSC KAMAZ. Moreover, in all variants of the charge amount of siluminum shavings AK18 was maintained at the level of 40%. The impact of the following factors characterizing features of the charge composition was investigated:

x_1 is a quantity of the mixture of aluminum shavings and ammonium nitrate in a relative mass ratio 1:3,5;

x_2 is an amount of graphite shavings;

x_3 is an amount of powdered lime.

The influence of an input controlled factors on the magnitude of the output parameters implemented using the method of proposed experiment [10]. For the structure of the linear mathematical model fractional factorial experiment 2^{4-1} was used for four input factors and three parallel observations when generating the ratio $x_4 = x_1 \cdot x_2 \cdot x_3$.

Mathematical model is the following:

$$y_1 = 16,07 + 2,69x_1 + 0,13x_2 - 0,93x_3 + 1,32x_4 \quad (1)$$

and is justified due to the fact that the difference between the maximum and minimum average values of the output parameter y_1 is significant. This follows from the comparison of the experimental values of Student criterion ($t_{эк}$) with table (t_α) for the number of degrees of freedom f=4 and reliability $\alpha = 0,95$. The calculation of the experimental values of Student criterion showed that $t_{эк} = 11,42$. This value significantly exceeds $t_\alpha = 2,78$. This fact allows us to conclude that the mathematical model (1) is justified.

The calculation of the total variance of the output parameter, called dispersion reproducibility, showed that $S^2_{y_1} = 0,337$.

After checking the significance of equation regression coefficients (1), we come to the conclusion that the a_2 coefficient is not significant, as $t_{2эк} < t_\alpha$. Hence, the equation of the regression, which describes the relationship between the strength of the moulding mixture in tension and the controlled factors, takes the following form:

$$y_1 = 16,07 + 2,69x_1 - 0,93x_3 + 1,32x_4. \quad (2)$$

Using Fisher's test the model was tested for the hypothesis of adequacy (suitability). The regression equation (2) is converted into the following form, if you use the values of the factors named in (natural) units:

$$y_1 = -1,65 + 0,11x_1 - 0,06x_3 + 0,88x_4. \quad (3)$$

We will hold a similar statistical study regarding the output parameter y_2 gas permeability of sand.

$$y_2 = 291,4 - 0,35x_1 + 0,41x_3 - 12,56x_4. \quad (4)$$

Analysis of the regression equation (4) establishing the influence of input parameters on the strength of sand under tension shows that the most significant parameters is x_1 – a quantity of the mixture of aluminum shavings and ammonium nitrate. The influence of other parameters (x_3 – number of powdered lime and x_4 – content of the slag product of the SHS process in sand) is equivalent. The difference in qualitative result of the influence of the parameters x_1 and x_3 on the strength of the moulding mixture should be noted. It is connected, obviously, with the level of binding properties of the slag product of SHS process. Previous studies showed that the cementing properties of the specified product grow with growth of parameter *x1* and fall with the growth of parameter x_3. As the increase in both parameters contribute to the saturation of the slag phase Al_2O_3 in the first case and CaO in the second one, the conditions of their variation in the composition of the charge of SHS process are the best prerequisites for the relevant formation of monocalcium aluminate $CaO{\cdot}Al_2O_3$ (CA), which defines the binding properties of the slag product. Indicated mineral in reaction with water eventually leads to the formation of compounds such as dicalcium of hydroalumination ($2CaO{\cdot}Al_2O_3{\cdot}8H_2O$ (C_2AH_8)) and aluminum hydroxide ($Al(OH)_3$), crystallization of which leads to the formation of cement with high strength. The experimental results allow to conclude that, under the influence of increasing the parameter x_1 and decreasing the parameter x_2 the above-described processes are initiated that causes the increase of the binding properties of the slag product of SHS process and accordingly the strength of the investigated cement sand. The increase of the parameter x_4, i.e. the amount of slag product of the SHS process in the molding mixture, also contributes to the last fact. It should also be noted the negative impact of excess lime in the slag product to the level of the knit of existing properties and, accordingly, the strength of the sand due to the fact that the possible recrystallization of dicalcium of hydroalumination (C_2AH_8) tricalcium hydroly-mint (C_3AH_6), accompanied by a decrease in the volume of tumors by about 25-30% and the occurrence of harmful stresses in the cement stone, causing a decrease in its strength.

Analysis of the research results on the gas permeability of sand (y_2) showed that this output parameter is significantly affected by the same input parameters as in the case of strength, but with the opposite quality. The reason for such relationship (4), obviously, is features of the flow crystallization processes at the interaction of monocalcium aluminate (CA) with water with the formation of the high density cement stone. In addition, the gaps between the crystals of dicalcium hydroaluminate filled with aluminum hydroxide, which has a dense structure. Therefore, those factors that contribute to the increase of strength properties of cement sand, oppositely affect its permeability. Moreover, as the results showed, the most important impact was made by the input parameter x_4 which is the content of the slag product in the molding mixture. Other parameters (x_1 and x_4) are equal.

3. Conclusion

Thus, the research results allow to conclude acceptability of sand properties using it as cementing component of the slag product of the process of obtaining SHS-cast iron of dispersed waste. Received adequate models allow to adjust sand strength and gas permeability through the parameters characterizing the charge composition of the SHS process, and the number in moulding mixture of the slag product – efficient substitute of alumina cement.

Reference

[1] Berg P.P. Quality mold [Kachestvo litejnoj formy.], M.:Mashinostroenie, 1971.–291p.
[2] Berg P.P. Molding materials [Formovochnye materialy.], M.:Mashinostroenie, 1963.– 408 p.
[3] Hudzievich R., Gutovski V. Rapid hardening fluid cement mixture [Bystrotverdejushhie zhidkopodvizhnye cementnye smesi. // Li-tejnoe proizvodstvo], 1976, – №1, – pp.17-19.
[4] Sano S., Kinoshita K. The process of rapid hardening cement mixtures forms manufacture [Tehnologicheskij process izgotovlenija form iz bystrotverdejushhih cementnyh smesej. Doklad na 34-m Mezhdunarodnom kongresse litejshhikov]. Paris, 1967. – pp.135-141.
[5] Brechko A.A., Velikanov G.F. Moulding and core mixture with the desired properties [Formovochnye i sterzhnevye smesi s zadannymi svojstvami.], L.: Mashinostroenie, Leningr. otd-nie, 1982. – 216p.
[6] Merzhanov A. G. Self-propagating high-temperature synthesis. Physical chemistry. [Samorasprostranjajushhijsja vysokotemperaturnyj sintez. Fizicheskaja hi-mija./ pod red. akad. Ja.M.Kolotyrkina], M.: Himija, 1983. – 224 p.

[7] Bikulov R.A., Safronov G.N., Harisov L.R. SHS-cast iron of dispersed waste engineering [SVS-ferrosilidy iz dispersnyh othodov mashinostroenija // Litejshhik Rossii.], 2010. № 8. – pp.15-17.

[8] Perepelicin V.A., Rytvin V.M., Gil'varg S.I., Ignatenko V.G., Kutalov V.G., Arzamascev N.N. Self- decomposing ferrography aluminosilicate slag - polyfunctional technogenic raw materials [Samoraspadajushhijsja ferrohromovyj aljuminotermicheskij shlak – polifunkcional'noe tehnogennoe syr'jo // Novye ogneupory.], 2009. № 10. – pp. 9-14.

[9] Pashhenko A.A., Serbin V.P., Starchevskaja E.A. Knitting materials [Vjazhushhie materialy.], Kiev: Vysshaja shkola, 1975. – 443 p.

[10] Novik F.S., Arsov Ja.B. Optimization of process metals technology by the methods of experiment planning [Optimizacija processov tehnologii metallov metodami planiro-vanija jeksperimenta.], M.:Mashinostroenie: Sofija: Tehnika, 1980. – 304 p.

IOP Conf. Series: Materials Science and Engineering **86** (2015) 012024 doi:10.1088/1757-899X/86/1/012024

Not adiabatic temperature of combustion gas-air mixture

O V Iovleva

Kazan Federal University, 18 Kremlyovskaya str., Kazan 420008, Russian Federation

E-mail: ovi_b@rambler.ru

Abstract. In the present work was carried out thermodynamic analysis, the purpose of which was to determine the formula for calculating the temperature of combustion gas-air mixture in non-adiabatic conditions.

1. Introduction

Combustion temperature - one of the basic parameters used in engineering calculations associated with the construction of pulsatory combustion devices [1], [2]. In such systems, heat losses due to the burning occurs in non adiabatic conditions [3], [4], [5]. Therefore, the well-known expression for the adiabatic combustion temperature must be corrected. The purpose of this work - a formula to calculate the temperature of the combustion gas-air mixture in the non adiabatic conditions.

2. Basic relationships

Consider steady state combustion of the gaseous mixture is continuously supplied into the combustion zone [6]. The mixture was warmed, ignited, the decomposition of the chemical components to the elements forming the heat release and combustion products, which temperature is called the temperature of combustion. When burning a non-closed volume pressure is constant.

According to the first law of thermodynamics, the amount of heat flow communication:

$$Q = \Delta H = H_1 - H_0, \tag{1}$$

where H_0 - the total enthalpy of the system with an initial temperature T_0 (the initial state of the flow), H_1 - the total enthalpy of combustion products (final state of flux with temperature T_c).

Since Q - total differential, the way in which the process of combustion, it does not matter. Then, if the mixture, which had a temperature T_0 warmed up temperature T_c, then immediately converted into combustion products, the thermal effect of the reaction is the same as in the case of real - Hess law. Hence $H_1 = H_{0,1}$, where $H_{0,1}$ - the enthalpy of combustion of the mixture at a temperature.

In general, the heat from the flow of the mixture - is the calorific value of fuel Q_1, which is deducted from the heat losses to the environment Q_2 and dissociation of molecules Q_3.

$$Q = Q_1 - Q_2 - Q_3. \tag{2}$$

The enthalpy of the mixture consists of enthalpies of its components:

$$H = \sum_i v_i H_i = \sum_i \nu_i c_{p,i} T, \tag{3}$$

where v_i - the number of moles, H_i - molar enthalpy i - the first component, $c_{p,i}(T)$ - the molar heat capacity at constant pressure, the temperature-dependent. To simplify the calculations take the average value $\bar{c}_{p,i}$ - for the temperature range from T_0 to T_c.

With regard to (2), (3) (1) we obtain

$$T_c = T_0 + \frac{Q_1 - Q_2 - Q_3}{\sum_i v_i \bar{c}_{p,i}}. \quad (4)$$

If you do not take into account losses, assuming $Q_2 = Q_3 = 0$ we get the so-called adiabatic combustion temperature:

$$T_c^* = T_0 + \frac{Q_1}{\sum_i v_i \bar{c}_{p,i}}.$$

3. Derivation of the formula for calculating the combustion temperature

In practice, usually measured fuel flow rates G_f (m^3 / s) and the oxidant, usually air G_a, and the mixture is characterized by a coefficient of excess air α. Obviously, $v_i \bar{c}_{p,i} = G_i \bar{c}_i$ where v_i - the number of moles per time unit, $\bar{c}_i$ - the average heat capacity per unit volume i of the ith mixture component at constant pressure (J / m^{3}K). Then

$$\sum_i v_i \bar{c}_{p,i} = v_f \bar{c}_{p,f} + v_a \bar{c}_{p,a} = G_f \bar{c}_f + G_a \bar{c}_a.$$

With $T_c < 3 \cdot 10^3\,K$ losses on the dissociation of the molecules can be neglected $Q_3 = 0$. This condition is satisfied by combustion of hydrocarbons in air, as $T_{c,\max} \approx 2*10^3\,K$. Heat loss to the environment can be represented in the form $Q_2 = \varphi Q_1$ where the ratio φ depends on the conditions under which combustion takes place.

When $\alpha \geq 1$ complete combustion of fuel and allocated the maximum possible amount of heat $Q_1 = G_f Q_f$, wherein Q_f – the heat of reaction per unit volume of fuel.

In view of (5), formula (4) takes the form

$$T_c = T_0 + (1-\varphi) Q_f / \bar{c}, \text{ при } \alpha \geq 1, \quad (6)$$

where the heat capacity of the mixture has

$$\bar{c}(\alpha) = \bar{c}_f + \varepsilon\alpha \bar{c}_a. \quad (7)$$

When $\alpha < 1$ the amount of fuel burned $V_f^* = V_f V_a / V_a^* = \alpha V_f$, then

$$T_c = T_0 + (1-\varphi)\alpha Q_f / \bar{c} \text{ when } \alpha \leq 1 \quad (8)$$

$\alpha = 1$ In formula (6) and (8) converge.

If the fuel - gas mixture to the volume fractions of known b_i, the flow rate of each component $G_i = b_i G_f$. If the fuel is specified weight fractions, whereas

$$G_i / G_f = \rho_f m_i / \rho_i m_f, \quad (9)$$

where m_i / m_f - the weight fraction, ρ_i - i-th component density, ρ_f - density fuel.

Heat of reaction corresponds to 1m^3 of fuel, then

$$Q_f = \sum_i b_i Q_{f,i}, \quad (10)$$

where $Q_{f,i}$ is the heat of reaction per unit volume i-th component.

Enthalpy of fuel

$$H_f = \bar{c}_f V_f T = \sum_i H_i = \sum_i \bar{c}_i V_i T = V_f T \sum_i b_i \bar{c}_i .$$

It follows that the heat of the fuel

$$c_f = \sum_i b_i \bar{c}_i , \tag{11}$$

where $\bar{c}_i$ is the heat capacity per unit volume of i -th component of fuel. Taking into account (7), (11) the heat capacity of the mixture of fuel, consisting of several gases with air equal

$$\bar{c}(\alpha) = \sum_i b_i \bar{c}_i + \alpha \varepsilon \bar{c}_a . \tag{12}$$

According to the formula(6), (7), (9), (12) to calculate the combustion temperature, it is necessary to know the heat loss factor φ.

The emissivity of the gas flame is small, because the radiation of the flame front is small. Visible radiation is behind the front in the combustion products containing carbon dioxide and water vapor. Measurements for open flames and inside the chambers with additional external heating of the walls showed that the combustion temperature is insensitive to changes in external conditions. Then the loss factor can be taken as a constant for a given gas, the independent and the conditions of heat transfer in the environment.

If you use the known value of the temperature of combustion to a certain value α, then from formulas (6), (7) it is easy to determine the loss factor and use it in calculations.

4. Conclusions

The result of the calculation for stoichiometric propane-air mixture is the following: φ=0,238. This value was taken to calculate the temperature of combustion of methane, ethane and propane with air in the range of $0.6 \leq \alpha \leq 1,4$. The obtained results nearly coincide with the known experimental data [7].

The resulting formula has been used in studies of pulsating combustion of propane-air mixtures [8]. The results of calculations of the excitation conditions of the gas oscillations and their frequency are in good agreement with the experimental data.

The work is performed according to the Russian Government Program of Competitive Growth of Kazan Federal University.

References

[1] Larionov V M and Nazarenko T I 2000 *Izvestiya Vysshikh Uchebnykh Zavedenij. Aviatsionnaya Tekhnika* **(4)** 69
[2] Larionov V M and Beloded O V 2003 *Izvestiya Vysshikh Uchebnykh Zavedenij. Aviatsionnaya Tekhnika* **(4)** 48-52
[3] Ermakov R A, Galiullin R G, Larionov V M, Nikolaev A N 2008 *Chemical and Petroleum Engineering* **44** 387-392
[4] Beloded O V and Larionov V M 2006 *Russian Aeronautics* **49(1)** 44-48
[5] Iovleva O V and Larionov V M 2007 *Russian Aeronautics* **50(3)** 303-308
[6] Zeldovich Y B, Barenblatt G I, Librovich V B and Makhviladze G M 1980 *M: Science* The mathematical theory of combustion and explosion 478
[7] Lewis B and Von Elbe G 1968 *Academic press inc. New York and London*
[8] Yallina E V , Larionov V M, Iovleva O V 2013 *Journal of Physics: Conference Series* **479(1)** 012017

IOP Conf. Series: Materials Science and Engineering **86** (2015) 012025 doi:10.1088/1757-899X/86/1/012025

Equation of fluctuations frequency of gas in the devices such as capacity-tube

O V Iovleva, E V Semenova

Kazan Federal University, 18 Kremlyovskaya str., Kazan 420008, Russian Federation

E-mail: techphys@kpfu.ru

Abstract. In this paper highlights gas oscillations in the devices consisting of a combustion chamber and a narrow tube for the outflow of combustion products. Considering the sound velocity gradient in the tube. The resulting equation of fluctuations frequency of gas in the test device with any geometrical parameters.

1. Introduction

Energy approach is often used to search the process of self-excitation of oscillations of the gas in the combustion chambers of power plants [1, 2]. This approach is possible if you know the frequency fluctuations of gas [3, 4].

In practice widespread device, which can be represented by a schematic diagram, comprising a combustion chamber 1, chimney flue gas 2, an input device 3 for supplying fuel and air (Figure 1). Gas enters the combustion chamber through holes, which the total area S_0 . Impedance input device Z0,0 depends on the acoustic properties of the supply system.

2. The equation of oscillation frequencies gas

The combustion chamber has acoustic properties of the container if its transverse size is much greater than diameter of the tube, and the longitudinal far less than wavelength. In this case, the motion of the gas in the tank can be ignored and assumed that changes in pressure at all points of the capacitance is the same, i.e.:

$$p'_0 = p'_V = p'_{2,0}, \qquad (1)$$

wherein p'_0 – acoustic pressure at the inlet to the capacity, p'_V – in capacity, $p'_{2,0}$ – inlet tube.

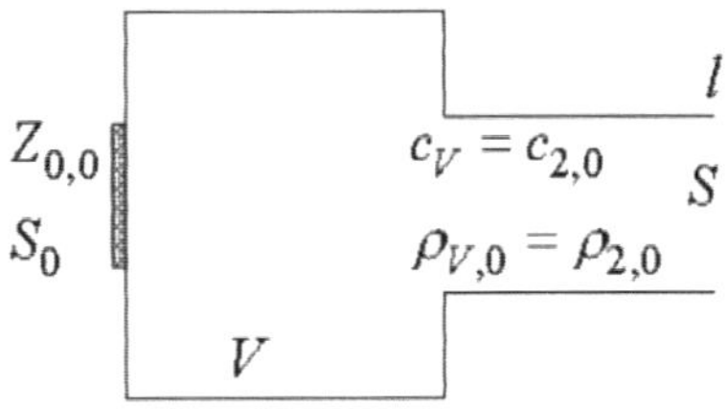

Figure 1. Device type capacity-tube.

The linearized mass conservation equation for the gas in the capacity has the form:

$$-\frac{V}{\rho_0 c_0^2}\cdot\frac{\partial p'_V}{\partial t}+S_0u'_0=S_cu'_{2,0},$$

wherein r_0, c_0 – the average gas density and velocity of sound, respectively, u'_0, $u'_{2,0}$ – pulsation of gas speed inlet and outlet capacity, respectively.

For harmonic oscillations $\partial p'_V/\partial t = i\omega p'_V$. After dividing by p'V, given the equation (1), we obtain:

$$\frac{S_cu'_{2,0}}{p'_{2,0}}=\frac{S_0u'_0}{p'_0}-\frac{i\omega V}{\rho_0 c_0^2} \tag{2}$$

At the tube inlet gas temperature is the same as in the capacity, then because of the cooling effect of the walls is reduced. Using the known expression [5-8] for the fluctuations of velocity and the gas pressure obtained for the linear distribution of the speed of sound. At the tube inlet by setting $x = 0$, we obtain:

$$u'_{2,0}=C_2e^{i\omega t}\cos\varphi_2,$$

$$p'_{2,0}=-i\rho_{V,0}c_{2,0}C_2e^{i\omega t}(\frac{b_2}{2\omega}\cos\varphi_2+\beta_2\sin\varphi_2).$$

wherein b_2 – gradient of the speed of sound, $\beta_2=(1-(b_2/2\omega)^2)^{1/2}$, φ_2 – phase angle, which depends on the boundary condition at the outlet pipe.

At the entrance into the capacity $p'_0=-iY_{0,0}u'_0$, где $Y_{0,0}$–the imaginary part of the impedance of the input device. In view of these remarks the expression (2) after some transformations takes the form:

$$\frac{b_2}{2\omega}+\beta_2 tg\varphi_2+F^{-1}=0, \tag{3}$$

where

$$F=(\frac{\omega V}{c_0S_c}-\frac{\varepsilon_0}{\overline{\overline{Y_{0,0}}}})\frac{c_0}{c_{2,0}},\qquad \varepsilon_0=\frac{S_0}{S_c},\qquad \overline{Y_{0,0}}=\frac{Y_{0,0}}{\rho_0c_0}.$$

Equation (3) allows calculate oscillation frequency of the gas under all conditions of the input and output devices of the type.

Without of a temperature gradient in the gas tube, opened on the outlet, when the inlet device has a large acoustic impedance, it is easy to obtain known equation of the vibration frequencies in a homogeneous gas tube with capacity:

$$tg(\frac{\omega l_c}{c_{2,0}})=\frac{c_{2,0}S_c}{\omega V}.$$

If the tube length is small compared with the wavelength, we obtain the well-known equation of the oscillation frequency of the Helmholtz resonator

$$\omega=c_{2,0}(\frac{S_c}{l_cV})^{1/2}. \tag{4}$$

If atmospheric air enters the combustion chamber through holes or via a short tube length l_0, which cross-sectional area is equal to the area of all holes, we have:

$$F=\frac{\omega V}{c_{2,0}S_c}-\frac{\varepsilon_0c_0^2}{\omega l_0c_{2,0}}.$$

If the tube length is so small that can be neglected sound velocity gradient, then $b_2=0$, $\beta_2=1$, $\varphi_2=-\omega l_c/c_{2,0}$. Then from equation (3) we obtain the formula:

$$\omega = (\frac{S_c}{l_c V})^{1/2}(1+\frac{\varepsilon_0 c_0^2 l_c}{c_{2,0}^2 l_0})^{1/2}.$$

Comparison with the expression (4) for conventional Helmholtz resonator, consisting of tube and capacity, shows us that the presence of holes or tube increases the oscillation frequency of gas.

3. Conclusions

Equation (3) makes it possible to calculate the frequency of oscillations of the gas in any type of device capacity-tube with predetermined geometric parameters. Substituting the calculated frequencies in the ratio corresponding to the energy method, will determine the conditions for self-excitation of oscillations of the gas in the test device.

References

[1] Larionov V M and Nazarenko T I 2000 *J. Izvestiya Vysshikh Uchebnykh Zavedenij. Aviatsionnaya Tekhnika* **4** 69
[2] Larionov V M, Beloded O V, 2003 *J. Izvestiya Vysshikh Uchebnykh Zavedenij. Aviatsionnaya Tekhnika* **4** 48-52
[3] Beloded O V, Larionov V M 2006 *Russian Aeronautics* **49** 44-48
[4] Iovleva, O V, Larionov V M 2007 *Russian Aeronautics* **50** 303-308
[5] Ermakov R A, Galiullin R G, Larionov V M, Nikolaev A.N. 2008 *Chemical and Petroleum Engineering* **44** 387-392
[6] Ermakov R A, Sadykov A F, Galiullin R G, Larionov V M 2011 *Chemical and Petroleum Engineering* **46** 749-751
[7] Semenova E V, Larionov V M, Kazakova E I 2014 *Journal of Physics: Conference Series* **567** 012033
[8] Fadeev S A, Kashapov N F and Larionov V M 2014 *IOP Conf. Ser.: Mater. Sci. Eng.* **69** 012006

IOP Conf. Series: Materials Science and Engineering **86** (2015) 012026 doi:10.1088/1757-899X/86/1/012026

Polishing and deburring of machine parts in plasma of glow discharge between solid and liquid electrodes

R I Valiev, A A Khafizov, Yu I Shakirov, A N Sushchikova

Kazan Federal Unuversity, Naberezhnye Chelny Institute
423810, Naberezhnye Chelny, Mira av., 68/19, Russia.

E-mail: rivaliev87@mail.ru

Abstract. The most effective way of materials processing in economical and ecological way is utilizing of the low temperature plasma. Modes of operation, in which the speed of polishing was increased in more than 2 – 2,5 times in comparison to electrical erosion method, were found. Roughness was about 0,16 – 0,08 micrometers.

1. Introduction

Cleaning [1, 2] and deburring of surfaces is an important technological operation of many industries. These technologies save energy, reagents, raw materials, time, increase productivity and product quality.

Of great interest are the processes in nonequilibrium discharge plasma [3-6], in which the temperature of the atoms and molecules is close to the ambient temperature, while the electrons have enough energy for excitation, dissociation and ionization of atoms and molecules. The use of such plasma improves the efficiency of many technological processes, such as plasma-chemical deposition of polymer and inorganic films, plasma anodizing, chemical synthesis, etc.

2. The experimental device

To streamline the process of cleaning and deburring, experimental plasma electro thermal installation with liquid electrode was developed [7-12] (Fig. 1). Bath is made of copper plates and has a cooling jacket. It is filled with the investigated electrolytes required concentration. Bath is connected to one terminal of the power source. The upper solid electrode is attached to the other terminal of the source. This electrode is installed on the coordinate device that allows to adjust the interelectrode distance and the coordinates X, Y. the electrolyte used was NaCl, $CuSO_4$, NH_4NO_3.

The experimental set-up [13] is designed to study the electrical discharge in a range of settings U_p =0,3 – 3000 V, current I=0,01÷200 A, $j_э$=0,1 - 25 A/cm^2, the interelectrode distance l = 0.1 to 100 mm. Cooling system installation serves for cooling the electrolyte. It is fed from the main water supply network and supply of electrolyte is made from a special tank.

[1] To whom any correspondence should be addressed.

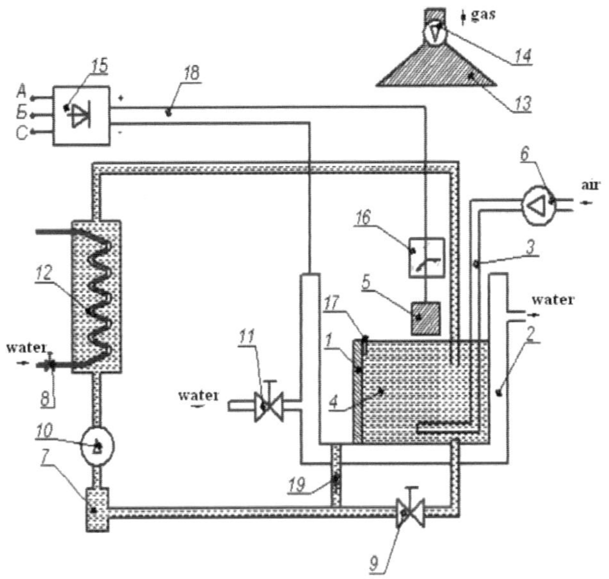

Figure 1. Functional diagram of the experimental installation
1 - electrolyte bath; 2 - jacket water cooling; 3 - bubbler; 4 - electrolyte; 5 - item; 6 - air compressor; 7 - filter; 8, 9, 11 - shut-off valve; 10 - pump; 12 - tank heat exchanger; 13-extractor hood; 14-exhaust fan; 15 - power supply; 16 - axis scanner; 17-thermometer; 18-conductors; 19 - return valve.

3. Results and their discussion

For the formation of a solid gas-vapor membrane [14], albeit unstable, it is necessary to provide energy in the anode area, sufficient to boil the electrolyte to some extent.

The highest quality of surfaces cleaning (Fig. 2) is carried out, if in the NaCl add boric or citric acid, which reduces sumelocenna electrolyte. It is especially important that the pH not only affects the anodic current density j_a, but also on the parameter R_a of the surface roughness.

Adding sodium nitrite to serve as a corrosion inhibitor and after treatment, the treated surface is corrosion resistant. Discharge voltage (≈ 120 V) has virtually no effect on the surface quality parameters and the details cleaning rate. The density of the same anode current affects the quality of treatment and duration of treatment. For example, for 20 seconds of current machining, with density of 1 A/cm2 at the depth $h = 5 \cdot 10^{-3}$ m, we obtained the surface roughness of 8th grade. The increase in current density up to 3 A/cm2 (at the time of processing 15s) has allowed to improve the quality of surface cleaning. A further increase in current density or the processing time almost does not affect the class of the surface roughness. The temperature of the electrolyte affects the quality of the processed surface. Studies have shown that with increasing temperature of the electrolyte decreases the performance of the process. So we cooled electrolyte using running tap water.

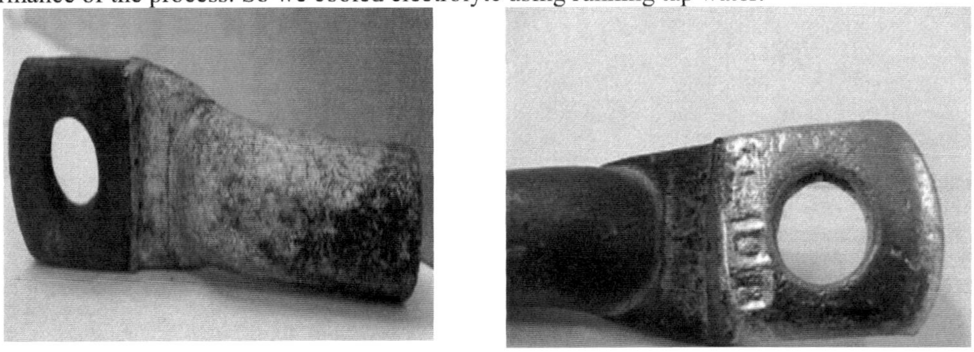

a) b)

Figure 2. Connection terminal:
a) before processing; b) after processing

The results are shown on (Fig. 3) of dependence studies of parameter R_z of the machined surfaces parts quality made of copper M1 and steel 25XHBA when processing discharge with liquid electrode (electrolyte temperature 300 K) depending on the current density on a solid electrode. It is seen that there is no noticeable improvement in the quality of the surface with increasing current density more than 3÷4 A/cm^2.

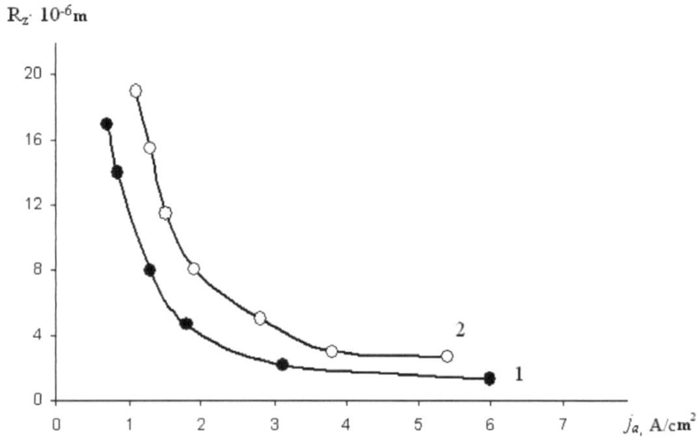

Figure 3. The quality of surface after treatment by plasma discharge with liquid electrode. The pressure P=105Pa, the temperature of the electrolyte $t_э$=300K, curve 1 - U = 120 V and the processing time t = 40 sec for copper M1, curve 2-magnetic core steel 25XHBA, U = 90V, t = 60 sec)

Methods of deburring and finishing increase the roughness class up to 8÷10, and diamond processing [15], paste polishing [16], electrochemical treatment [17] allow to achieve a roughness class of 13÷14. These methods are not always reliable, they are characterized by low productivity, complexity and irregularity of the removal of the edges, i.e. the surface geometrical characteristic forming process is difficult to manage.

Effective deburring (Fig. 4) is achieved remaining after the machining of complex configuration parts edges, an electric discharge between the workpiece and the liquid electrode. The area of the gas-vapor layer starts to expand with the speed of 0.005÷0.01 m/s. Current is growing, and in 8÷12 sec, under critical conditions, the layer is broken by the explosion, then the process repeats. With h =10^{-3}m, the surface area of the layer reaches 0,012÷0,5m^2, and the discharge current - 200A. Studies have shown that the process of removing the burrs should be carried out at 4 A/cm^2 < j_a <13A/cm^2, a voltage of 80V < U_p < 120 V.

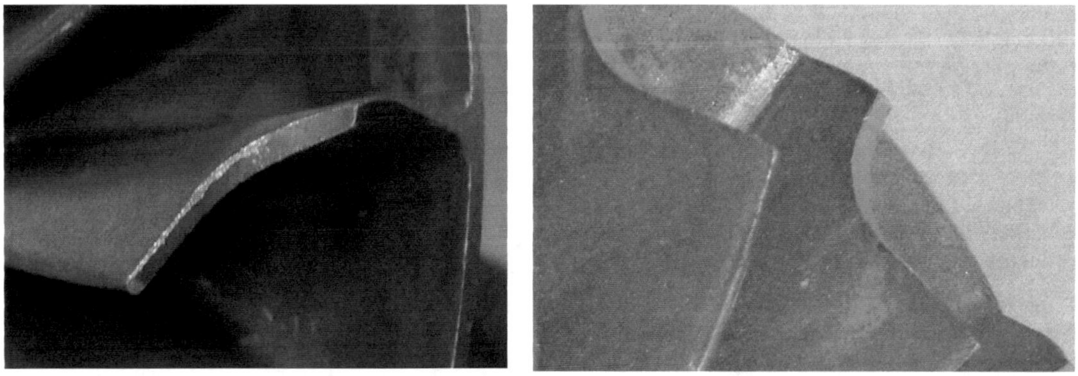

a) b)

Figure 4. Complex Configuration Part Processing
(a) before processing; b) after processing

As a result of processing samples weight change was observed. The average weight change of a copper sample when the duration of treatment is $t_{обр}$ = 25 seconds is Δ $P_м$ = 380 milligrams, for a sample of steel for the same processing time it is Δ $P_{ст}$ = 350 milligrams (Fig. 5)

The analysis of the above dependences has shown that the weight change of a copper samples and electrical steel processing is 27 seconds and is almost the same.

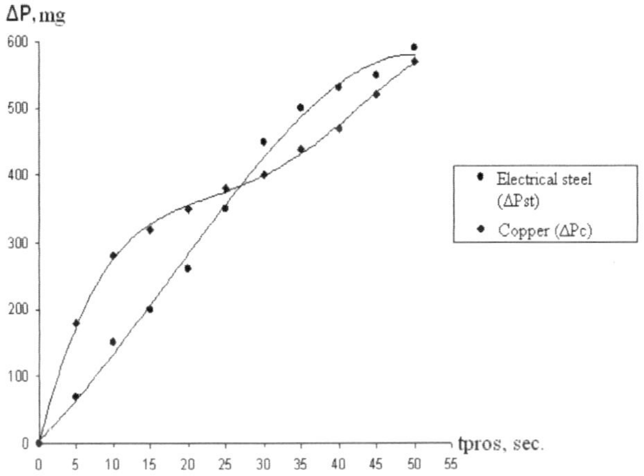

Figure 6. Mass change of steel and copper parts

Thus, provided that the weight change ΔP of the workpiece after processing does not exceed the permitted limit, taking into account the installation performance and to achieve the desired (requested) quality of surface treatment, the required length of time treatment will be 20 - 35 seconds.

4. Conclusions

For machining the rectified voltage U = 400 - 550 V power source is required with power up to 100 kW. The voltage discharge has virtually no effect on the surface quality parameters and the speed of cleaning components. Due to the analysis the conditions were established under which the deburring performance increases in 2 - 2.5 times in comparison with the electroerosive method, while the roughness is reduced to 0.16 - 0.08 micrometer (roughness class of 8 - 9).

References

[1] Smirnov N.S., Prostakov M.E. Cleaning the surface of the steel [Ochistka poverhnosti stali], M., 1978. – 230 p.

[2] Morozov A.P. i dr. Electric cleaning rod. Thermal engineering questions; application questions STP in metallurgy [Jelektrorazrjadnaja ochistka katanki. V sb. Teplotehnicheskie voprosy primenenija voprosy NTP v metallurgii]. Sverdlovsk, 1985.pp. 102-104.

[3] Tazmeev Bulat Harisovich. Electrical and thermal characteristics of generators nonequilibrium discharge plasma with liquid electrodes [Jelektricheskie i teplovye harakteristiki generatorov neravnovesnoj gazorazrjadnoj plazmy s zhidkimi jelektrodami: Dis. kand. tehn. nauk: 01.02.05[, Kazan', 2000- 173 p.

[4] Kashapov N F, Saifutdinov A I and Fadeev S A 2014 *J. Phys.: Conf. Series* **567** 012004

[5] Fadeev S A and Kashapov N F 2013 *J. Phys.: Conf. Series* **479** 012009

[6] Dautov G, Dautov I, Fayrushin I and Kashapov N 2013 *J. Phys.: Conf. Series* **479** 012014

[7] Shakirov Ju.I. Characteristics of plasma electrothermal installation with liquid cathode [Diss. na soiskanie uch. stepeni k.t.n. "Harakteristiki plazmennoj jelektrotermicheskoj ustanovki s zhidkim katodom".], Leningrad. 1990. -132 p.

[8] Fadeev S A, Kashapov N F and Larionov V M 2014 *IOP Conf. Ser.: Mater. Sci. Eng.* **69** 012006

[9] Gajsin F.M., Son Je.E. Shakirov Ju.I. Volumetric discharge in a vapor-gas medium between the

solid and liquid electrodes [Obemnyj razrjad v parogazovoj srede mezhdu tverdym i zhidkim jelektrodami], Moskva 1990.

[10] Gajsin F.M., Shakirov Ju.I., Hakimov R.G. Study of discharge between the solid and liquid electrodes [Issledovanie razrjada mezhdu tverdym i zhidkim jelektrodami // Tezisy dokladov Respublikanskoj nauchno-tehnicheskoj konferencii] KamAZ-KamPI, Nab. Chelny 1990.- p.161.

[11] Kashapov L, Kashapov N and Kashapov R 2013 *Journal of Physics : Conference Series.* **479** 012005

[12] Valiev R.I.,Shakirov Ju.I., Il'in V.I., Shakirov B.Ju. The process control system of surface treatment products plasma electrothermal installation with liquid electrode [Sistema upravlenija processom obrabotki poverhnosti izdelij plazmennoj jelektrotermicheskoj ustanovkoj s zhidkim jelektrodom // Nauchno - tehnicheskij vestnik Povolzh'ja]. Kazan', 2012, №1, pp.131-138.

[13] Shakirov Ju.I., Valiev R.I., Hafizov A.A., Shakirova G.Ju. Multi-channel plasma installation electrolytic cathode [Mnogokanal'naja plazmennaja ustanovka s jelektroliticheskim katodom // Nauchno-tehnicheskij zhurnal «Avtomobil'naja promyshlennost'»], Moskva, 2011, №2, pp. 36-38.

[14] Valiev R.I., Il'in V.I., Hafizov A.A., Shakirov B.Ju., Shakirov Ju.I. A mathematical model of the steam-shell electric discharge with liquid cathode [Matematicheskaja model' parogazovoj obolochki jelektricheskogo razrjada s zhidkim katodom // Nizkotemperaturnaja plazma v processah nanesenija funkcional'nyh pokrytij. V Respublikanskaja nauchno-tehnicheskaja konferencija: sbornik statej. Kazan', 4-7 nojabrja 2013.] Kazan': Otechestvo, 2014.pp.135-138.

[15] Dolgih A.I. Bench work: tutorial [Slesarnye raboty: Uchebnoe posobie], M.: Al'fa-M: INFRA-M, 2010. - 528 s.

[16] Ivanov I.S. Mechanical engineering: a manual [Tehnologija mashinostroenija: Ucheb. Posobie]- M.: INFRA-M, 2009. - 192 p.

[17] Adaskin A.M., Zuev V.M. Science and technology of materials [Materialovedenie i tehnologija materialov] M: Forum, 2010.-336 p.

IOP Conf. Series: Materials Science and Engineering **86** (2015) 012027 doi:10.1088/1757-899X/86/1/012027

The choice of equipment for automation of hazardous production facilities

R R Ziyatdinov
Kazan Federal University, Naberezhnye Chelny Institute
423810, Naberezhnye Chelny, Mira av., 68/19, Russia.

E-mail: chelny@kpfu.ru

Abstract. The characteristics that must be considered when selecting electrical equipment used at hazardous production facilities are discussed. To minimize hardware costs and security features it is suggested to use expert system based on the productional model.

1. Introduction

A large share of enterprises in various sectors of the economy refers to dangerous industrial objects. Most of them belong to the dangerous industrial objects from the point of view of explosion produced, used, processed, stored, transported and slaughtered flammable and combustible substances. The range of such enterprises is much wider than it seems. In addition to the classic dangerous industrial objects, such as coal mines, oil refineries, pumping stations, gas and oil objects and so on, it also includes, for example, painting chambers of machine-building enterprises, flammable liquids warehouses (flammable liquids) and even granary.

Automation of such objects must be secured from the standpoint of eliminating the potential of applied electrical equipment to become a source of initiation of explosion or fire. However, it should be noted that the higher the level of explosion-proof electrical equipment, the more expensive it is. Therefore, in the general case, the task of selecting the equipment is reduced to the minimization of the cost of equipment subject to constraints imposed by the need of security.

2. Classification of hazardous areas

Classification of hazardous areas is used for equipment selection by its protection level, ensuring safe operation of such equipment in the appropriate hazardous area. Dangerous industrial objects can have explosive zones of different classes, including explosion-proof zone. The class of hazardous areas is determined by the technologists together with the specialists of the design or operating organization. According to GOST R 51330.9-99 (IEC 60079-10-95) classification of hazardous areas should be undertaken by personnel who are familiar with the properties of combustible gases and vapors, knowledgeable technological process and equipment, in cooperation with safety engineers, electricians and other technical personnel [1].

Nowadays in the Russian Federation hardware for operation in explosive environments requirements are established by Technical regulations of the Customs Union "On safety equipment for working in hazardous environments" (TR TS 012/2011), which entered into force on February 15, 2013.

According to TP TC 012/2011 for explosive gas atmospheres (mixtures of flammable gases or vapours with air) hazardous zones are divided into three classes (see Fig.1):

- zone of class 0 is an area in which an explosive gas mixture is present continuously or for long periods of time;
- zone of class 1 is an area in which there is a probability of explosive gas mixtures under normal operating conditions;
- zone of class 2 is an area in which there is an unlikely presence of explosive gas mixtures under normal operating conditions, and if it does present, then it takes place rarely, and it is for a very short period of time.

For explosive dust environments the following classes of hazardous zone present:

- zone of class 20 is the area in which combustible dust, as a cloud, is present continuously or partially under normal operation in quantities capable of producing a concentration sufficient to explosion of combustible or inflammable dust / air mixtures, and/or where layers of dust arbitrary or excessive thickness can be formed;
- zone of class 21 is the area not classified as zone class 20, in which combustible dust is in the form of clouds may not be present under normal operation in quantities capable of producing a concentration sufficient to explosion of combustible dust in mixture with air;
- zone of class 22 is the area not classified as zone class 21, in which a cloud of combustible dust may occur rarely and is only stored for a short period or in which the accumulation of layers of combustible dust may occur when an abnormal operation mode, which can lead to the emergence able to ignite mixtures of dust in the air. If on the basis of abnormal conditions the elimination of savings or dust layers cannot be guaranteed, then the area is classified as zone class 21.

Separate classification of zones for gases and dust is explained by the fact that unlike zones for gas or vapors, flammable liquids, areas prone to ignition of combustible dust may be classified depending on the normal or emergency conditions and from time to time. Enhanced ventilation can cause clouds of dust and, respectively, increase but not reduce the risk [2].

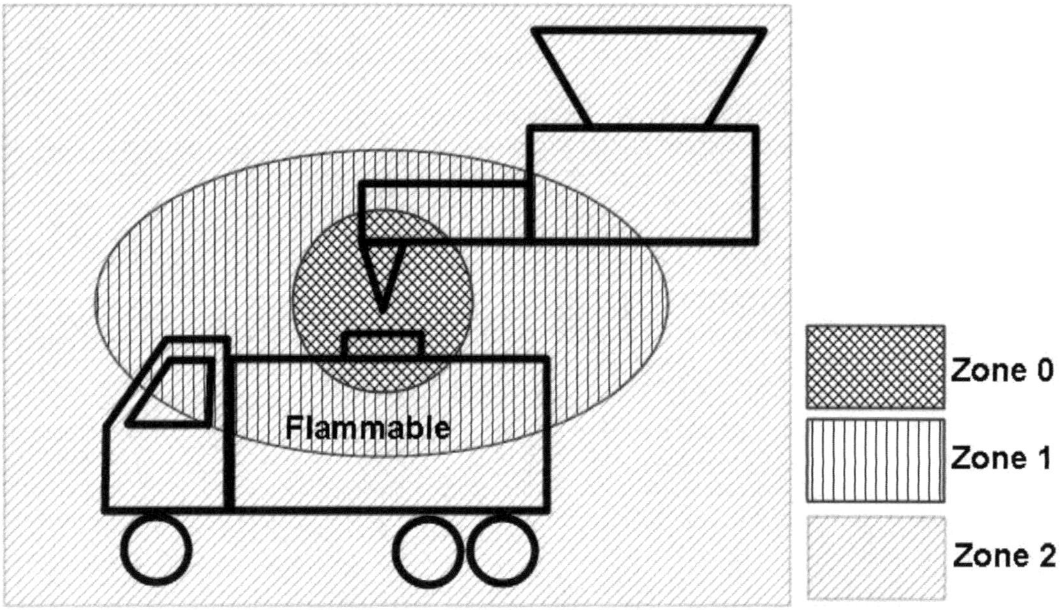

Figure 1. An example of hazardous zones distribution

Nowadays in Russia there are several series of standards and rules governing the use of explosion-proof equipment. For example, according to Chapter 7.3 "Rules for electrical installation" (REI) there is a different classification of hazardous areas for mixtures of gases and vapours with air:

• zone of class B-I is a zone located in the premises in which flammable gases and vapours of flammable liquids are emitted in such amount and with such properties that they can form with air an explosive mixture during normal operation modes, for example, when loading or unloading technological devices;

• zone of class B-Ia is a zone located in the room in which under normal operation a potentially explosive mixtures of flammable gases or flammable liquids or steam with air are not formed, but possible only during accidents or malfunctions;

• zone of class B-Ib is a zone located in the room in which under normal operation a potentially explosive mixtures of flammable gases or vapors, flammable liquids with air are not formed, but possible only during accidents or malfunctions, and which have some special features listed in paragraph (7.3.42 REI);

• zone of class B-Ig are spaces in outdoor installations: process units containing flammable gases or flammable liquids, ramps for unloading and loading of flammable liquids, etc.

Thus, for explosive mixtures of gases or vapours with air TR TC 012/2011 establishes three classes of zones, and REI – four, so formal correspondence between the two classifications cannot be set.

It should be noted that zone of class B-1a and B-1b can not be defined as zone 2, since the possibility of occurrence of failure from a legal perspective is not defined as frequency of occurrence and duration of an explosive mixture. However, for zones of class B-1a and B-1b it is necessary to use equipment designed for use at least in the zone 2 (equipment reliability against explosion).

To avoid error in the determination of compliance zones, equipment for zones of class B-Ig must have class explosion-proof corresponding to zone 1 (explosion proof equipment).

In zone B-I, in which an explosive gas mixture is present continuously or for long periods of time, it is possible to use equipment designed for operation only in zone 0 (explosion-proof equipment).

3. The choice of equipment

Explosion-proof electrical equipment is divided into groups:

I – equipment intended for use in underground mines and their surface structures, hazardous mine gas and/or combustible dust (mine explosion-proof electrical equipment). Depending on the design, the equipment of group I can have one of three levels of protection.

II – equipment intended for use in places (except underground mines and surface structures), with hazardous explosive gas environments. Depending on the design, the equipment of group II may have one of three levels of protection. Equipment of group II may be divided into subgroups depending on the category of explosive mixture, for which it is intended (IIA, IIB, IIC). For example, the subgroup IIA include methane, propane, subgroup IIB – ethylene, subgroup IIC – hydrogen, acetylene.

III – equipment intended for use in places (except underground mines and surface structures) with hazardous explosive dust environments. Depending on the design, the equipment of group III can have one of three levels of protection. Equipment of group III may be divided into subgroups depending on the characteristics of the explosive atmosphere for which it is intended (IIIA, IIIB, IIIC).

Explosion-proof electrical equipment, depending on the danger of becoming a source of ignition and conditions of its application in explosive environments, can be classified by levels of protection:

1) especially explosion-proof equipment (very high level);
2) explosion-proof equipment (high level);
3) equipment of high reliability against explosion (advanced level).

Depending on the maximum surface temperature, electrical equipment of group II is divided into temperature classes:

1) T1 – 450 °C;
2) T2 – 300 °C;
3) T3 – 200 °C;
4) T4 – 135 °C;
5) T5 – 100 °C;
6) T6 – 85 °C.

Depending on the envisaged special measures to prevent the ignition of a surrounding environment, the equipment may have one or a combination of several types of protection, for example:

• Explosion protection of type “i” with levels “ia”, “ib”, “ic” (spark-safe circuit);

• Explosion protection of type “d” (flameproof enclosure);

• Explosion protection of type “s” (special kind of protection);

etc.

Example of marking of explosion-proof electrical equipment:

0ExiaIIBT3 – especially explosion-proof equipment with type of protection spark-safe circuit of level "ia", for explosive gas group IIB with an ignition temperature of not less than 200 ºC.

The choice of equipment for automation of dangerous industrial objects (from the point of view of explosion) should include the following steps:

1) Defining the group of electrical equipment depending on the scope (I, II or III).

2) Defining the class of hazardous zones in accordance with technical passport of a hazardous production facility, project or operational documentation. Depending on the class of hazardous areas and conditions of use of the equipment, the level of protection equipment is chosen.

3) Identifying subgroups of equipment depending on the energy of the ignition of explosive atmospheres (explosive composition of the mixture) in the labeling follows the type of protection;

4) Determining the type of protection equipment. It is necessary to consider the following:

• different types of protection are characterized by different costs of operation;

• not all levels of protection can be implemented by any type of protection. So, for example, especially explosion-proof equipment may not be realized by the application of protection type "flameproof enclosure" (protection type "d")

5) Determining the temperature class of the equipment (marking follows a group of electrical equipment) according to the temperature of ignition of an explosive mixture.

The choice of the necessary equipment for the automation of a hazardous production facility requires highly skilled personnel, because the wrong choice can lead to incidents and accidents. One of the solutions to simplify the choice of equipment is the use of an expert system based on the production model. At each step rule-based models are used, allowing to represent knowledge in the form of "If (condition), then (action)". Under "condition" some sentence-pattern is realized in accordance to which the search in the knowledge base is carried out, and under "action" the action is realized at the successful outcome of the search [3].

Example of rules usage when choosing equipment:

If the equipment is intended for use in places (except underground mines and surface structures), hazardous explosive gas environments, and if the equipment is intended for use in zone 0, and if an explosive mixture is classified as IIB, and if the explosion protection of the equipment necessary to provide a form of "spark-safe circuit", and if the auto-ignition temperature explosive mixture is at least 200 ° C and 300 ° C, then the equipment must be selected by marking 0ExiaIIBT3.

4.Conclusion

Application of expert system in the selection of electrical equipment for use in hazardous production facilities will allow, on the one hand, to reduce the risk of accidents due to the exclusion of incorrect equipment selection, on the other hand, to optimize the cost of the equipment used.

References

[1] Denisenko V. The choice of automation hardware for dangerous industrial objects [Vybor apparatnyh sredstv avtomatizacii opasnyh promyshlennyh ob’ektov // Sovremennye tehnologii avtomatizacii], 2005. – №4. – pp.86-90.

[2] Classification of hazardous areas and marking of explosion-proof equipment [Klassifikacija vzryvoopasnyh zon i markirovka vzryvozashhishhennogo oborudovanija [Jelektronnyj resurs]. – available at: http://www.cortemgroup.ru/cortemqfm000r08u80040b.html.

[3] Information systems in metallurgy: lecture Notes (individual chapters from the textbook for

IOP Conf. Series: Materials Science and Engineering **86** (2015) 012027 doi:10.1088/1757-899X/86/1/012027

high schools) [Informacionnye sistemy v metallurgii: Konspekt lekcij (otdel'nye glavy iz uchebnika dlja vuzov)], Ekaterinburg: Ural'skij gosudarstvennyj tehnicheskij universitet, UPI, 2004. – 495 p.

IOP Conf. Series: Materials Science and Engineering **86** (2015) 012028 doi:10.1088/1757-899X/86/1/012028

Algorithm and data support of traffic congestion forecasting in the controlled transport

S V Dmitriev

Kazan Federal University, Naberezhnye Chelny Institute
423810, Naberezhnye Chelny, Mira av., 68/19, Russia.

E-mail: dmitriev37@yandex.ru

Abstract. The topicality of problem of the traffic congestion forecasting in the logistic systems of product movement highways is considered. The concepts: the controlled territory, the highway occupancy by vehicles, the parking and the controlled territory are introduced. Technical realizability of organizing the necessary flow of information on the state of the transport system for its regulation has been marked. Sequence of practical implementation of the solution is given. An algorithm for predicting traffic congestion in the controlled transport system is suggested.
Key words: Control, area, filling, transport, highway, parking, intensity, flow, congestion, forecast technical, facilities, algorithm.

1. Introduction

Logistical approach to the formation of a concentration-distribution systems of product movement allows us to justify economically suitable boundaries of geographical location of the of zones of potential sales of products and services for manufacturers of products and enterprises of commodity-distribution network [5,6,7]. However, with the implementation of projects it is possible to observe a manifestation of the influence of factors not covered by the computational methods of logistic systems.

The continuous growth of the global vehicle fleet leads to the fact that traffic jams are becoming more and more likely, with a steady trend towards increasing of economic losses caused by their appearance. In this regard, the task of analyzing the conditions of emergence, development and evolution of the traffic congestion is urgent and that will help to predict and prevent their formation. To solve this problem it is necessary to identify the main factors and develop mathematical models of traffic processes, to investigate the mechanisms of formation, "life" and the elimination of traffic congestion, the impact of internal and external factors on these processes. Similar systemic tasks are worked out in relation to the organization of work of automated multiproduct manufacturing site of engineering enterprise, which can be adapted to the problems of traffic management [1,2,3,4,8]. All that will provide the reliable forecasts of the traffic congestions emergence of vehicles in order to prevent them. Despite the lack of system solutions of these tasks, and the results of the study of influencing factors, it is possible to offer basic approaches to solving of common problem at the level of its data support and algorithmization. In this paper we consider all necessary and possible sources of information about the elements of the road traffic and algorithms for information processing for decision-making on the prevention of traffic congestion.

2. Basic part

Analysis of publications on this topic [9,10,13] shows that in different countries, including the Russian Federation, serious attention is paid to monitoring and regulation of road transport processes, and the Moscow City Government conducts the systematic work aimed at technical and organizational support to reduce the danger of traffic congestion [11, 12]. Really in the world are used many varieties of traffic regulation means: traffic lights of varying complexity, video cameras as part of subsystems of video, information support of road users. A video camera mounted on a dangerous stretch of road translates the visual and numerical information in the information processing module, which analyzes the movement of vehicles and determines the specified integrated motion estimations. There are also used: multiposition road signs, light boards with interchangeable information about the traffic situation on the directions of movement, special radio and video channels. Radars and collision warning devices on the road; locking devices; satellite navigation and determining the location of the vehicle system; mounted in the roadway luminous beacons to determine the speed, identifying license plates of passing vehicles; video subsystem with an automatic fixing of a car number and driver's face with giving a ticket to owner of the car at excess speed; satellite tracking system for compliance with parking regulations; laser scanners for fast fixing of the place of traffic accident; automated system for collecting payments for the right to enter the city center during peak hours; self-learning traffic signal, automatically changing the mode of switching depending on the traffic on the streets at a crossroads – are developed.

Thus, there are technical opportunities for obtaining a variety of traffic information and the means to implement a centralized control within the controlled transport system. However, the likelihood of congestion formation on the roads is continuously increasing, which makes return to the tasks of their prevention. Consider a general statement of the problem. It is very easy. For a controlled area of the transport system is necessary to determine the conditions under which the likelihood of congestion becomes close to 100%.

Obviously, this is possible when the controlled area is filled with vehicles so that it becomes difficult to move them along the roads of the territory, and then comes the stop of their movement. High intensity of input vehicles streams contributes to filling the area and high intensity of output streams - to releasing. The intensity of the traffic flow on the road is determined with all other equal conditions by the width of the roadway and vehicle speed. Speed of the vehicle within the controlled area, in turn, is determined by the established limitations of motion in the form of road signs and traffic lights devices and the degree of cluttering a roadway with parked cars, snow, repair work, the consequences of the accident. And here it should be noted that the majority of traffic signs and control devices help to reduce the average speed of the vehicle along a controlled area, and most of the sanctions to drivers of vehicles are imposed for exceeding the speed prescribed by traffic signs regulation. This apparent systemic contradiction can be resolved only by the systemic consideration of the issue. It is also clear that the preferred solution would be directed to the possible increase a throughput of a controlled territory, including at the expense of increasing of the average speed of vehicles, the introduction of unregulated intersection of highways in different levels, and creating of underground and overground pedestrian crossings. All this can stabilize the potential throughput of a controlled territory at the highest level. However, a high throughput of a controlled territory in relation to the traffic flows can not be the only of its function.

Another important purpose of any controlled area is the ability to accept and place (park) vehicles, as even on the country roads may occur planned and forced stops, and settlements, especially large cities are the areas to where, in general, vehicles move transporting people and cargo for various purposes, thus , vehicles stoppings and parkings are an integral part of traffic and should also be designed on the basis of a systematic approach with the design of areas and transport routes. Considering the problem of the vehicle parking in terms of providing throughput of the controlled area, it should be noted that the presence of organized parking contributes to unloading the roads from vehicles and thereby enhances the throughput capacity of the controlled area. But most of parked vehicles after parking time joined the transport stream again, reducing the throughput capacity. Availability of technical devices parking control allows you to organize monitoring of the degree of

IOP Conf. Series: Materials Science and Engineering **86** (2015) 012028 doi:10.1088/1757-899X/86/1/012028

filling of parking by vehicles and consider these data in predicting the likelihood of congestion of vehicles.

Taking into account stated above, and considering the use of any available technical means, we can consider an algorithm predicting traffic congestion in the controlled transport system.

1. Number all input (i) and all output (j) transport highways of the controlled territory, including the parking ins and outs of vehicles.

2. With the help of statistical observation or calculation methods to determine the value of:

NкрA - critical value of the filling of the controlled territory by vehicles complying A% of probability of formation of congestion;

Nmax - maximum permissible value of the filling of the controlled territory by vehicles in which the probability of congestion A = 100%.

4. Define by monitoring of highways and parkings the values:

Nм0,Nc0 - respectively, the initial number of vehicles on the carriageway of highways and parkings of the controlled territory.

5. Perform using technical control devices a monitoring of the time streams intensity of the vehicles for each of i input highways n ini(τ) and for each of j output highways n outj (τ) of the controlled territory.

6. To determine the number of vehicles entering ((N ini)) the controlled territory and exiting ((N outj)) out of it for forecast period Δτ = τ к - τ н [c], wherein τ к, τ н - respectively, initial and final moments of time of the forecast period:

$$N\ ini = \int_{\Delta\tau} n\ ini(\tau)\ d\tau\ , \quad (1)$$

$$N\ out\ j = \int_{\Delta\tau} n\ outj\ (\tau) d\tau. \quad (2)$$

7. Determine the current value of the filling of the controlled territory by vehicles, summarizing the obtained values:

$$N = N_{м0} + N_{c0} + \Sigma_i\ N\ ini\ - \Sigma_j\ N\ outj \quad (3)$$

8. Compare the result at the end of the forecast period with the critical value of filling of the controlled territory by vehicles with the probability A % of the congestion formation

$$N \le N_{кpA} \quad (4)$$

9. When satisfying the condition (4) continue to monitor the intensity in the time streams of vehicles of the controlled territory by the transition to i. 5 of the algorithm.

10. In case of violation (4) make a notification of road users about load of highways, start a traffic control according to the unloading scheme of the considered highway of the controlled territory.

11. Compare value N, obtained on a parity (3), п.7 of an algorithm with a maximum value of filling of the controlled territory

$$N \le N_{max} \quad (5)$$

12. When condition (5) continue to monitor the intensity in the time of vehicle streams of the controlled territory, passing to i. 5 of the algorithm.

13. In case of violation (5) overlap with the input highways of the controlled transport system taking measures to alert road users on the near and distant approaches of all input highways.

14. These monitoring results with reference to real time and weather conditions direct to the precedent bank for further use in statistical studies and as a real precedent in the analysis of road situations in a controlled transport system.

The considered algorithm can be applied as a control procedure in any transportation system and can be adjusted according to the settings as they gain experience of its application. If necessary, the major transportation systems can be divided into separate fragments, working on the same algorithm with centralized coordination of their functioning.

3. Conclusion

1. A general statement of the problem of traffic congestion forecasting in the controlled transport system is formulated.

2. Systemic contradiction in the current system of regulating traffic flows is revealed.

3. An algorithm for predicting traffic congestion in the controlled transport system is designed.

References

[1] Adgamov RI, SV Dmitriev, Karimov TN, AK Khayrullin Mathematical model to evaluate the performance of automated multiproduct manufacturing in mechanical engineering // Nauchno-tekhnicheskii vestnik Povolzh'ya. - 2012, №6. P.68-71.

[2] Adgamov RI, SV Dmitriev, Karimov TN, AK Khayrullin Performance of automated manufacturing site of engineering enterprise with a centralized management system // Nauchno-tekhnicheskii vestnik Povolzh'ya . - 2012, №6. P.72-75.

[3] Adgamov RI, SV Dmitriev, Karimov TN, AK Khayrullin Evaluation of automated manufacturing site performance of engineering enterprise with a hot standby control system // Nauchno-tekhnicheskii vestnik Povolzh'ya . - 2012, №6. P.76-79.

[4] Adgamov RI, SV Dmitriev, Karimov TN, AK Khayrullin A mathematical model of an automated performance of multiproduct manufacturing in mechanical engineering, taking into account the interaction of internal factors // Nauchno-tekhnicheskii vestnik Povolzh'ya. - 2013, №1. P.75-78.

[5] Anikin BA, Tyapukhin AP Commercial Logistics: Textbook. - M .: Prospect, 2009. - 432 p.

[6] R. Valiev, Karimov TN, Sibgatullin RK, AH Khayrullin Multicriteria scheduling tasks of delivering fuel by energy and transport company to networks of gas stations // Nauchno-tekhnicheskii vestnik Povolzh'ya. - 2013, №1. P.143-145.

[7] Hadjinsky AM Logistics: Textbook. - 16 th ed. Revised. and add. - M .: Publishing and Trading Corporation "Dashkov & Co", 2008. - 484 p.

[8] Zubkov EV, Dmitriev SV, Khayrullin AH Algorithmization of processes of automated testing of diesel engines. Kazan: Kazan. University Press, 2011.

[9] Kasmynin A. When the whole world will stand. Phenomenology of traffic jams. - M .: "Zavtra» №49 (994). - 2012

[10] Kasmynin A. Again "jam"! Is it possible to solve the transportation problem in the Russian capital. - M .: "Zavtra» № 52 (997). - 2012.

[11] Liverovskii E. Managing of the traffic lights of the capital is now concentrated in the hands of Moscow Government [electronic resource] URL: http://maps.mail.ru/articles/?id=51 (date accessed 01/12/2014).

[12] On the transfer to public institutions of Moscow city - Centre of traffic organization of Moscow Government – city-wide automated traffic control system SYSTEM - "START" [electronic resource] URL: http://www.dtis.ru/Doc/PProjects/project_2010-11- 29.pdf (date accessed 12/01/2014).

[13] Orlov M. Kukushkin L. Tverskaya congestion. - M .: "Zavtra» № 49 (994). - 2012.

IOP Conf. Series: Materials Science and Engineering **86** (2015) 012029 doi:10.1088/1757-899X/86/1/012029

Building axiological competence of graduate students by means of project-based learning

S.I. Gilmanshina, I.R. Gilmanshin

Kazan (Volga Region) Federal University, Russia

E-mail: gilmanshina@yandex.ru

Abstract. The article defines the essence of axiological competence, objectives and pedagogical conditions of its formation among the students of graduate program in Biotechnology. The authors provide requirements for energy-saving technologies project-based learning and specify the application of the latter with the view of competence building.

1.Introduction

According to the modern university graduate competence model, a two-tier higher education (undergraduate and graduate studies) should be student-oriented. Along with subject knowledge and skills it is necessary to form personal qualities which are highly demanded by this professional field. Natural disasters of the last decade have proven conclusively that industry and agriculture should not be the only sectors developed in terms of environmental protection. It is essential to organize the educational process on the basis of the fundamental unity of its scientific and humanitarian components. The latter implies the development of axiological competence, ability to harmonize relations in the system "man-nature-society", which is vital for graduate students in Biotechnology.

At the same time, project-based learning has recently been widely used in education. Its didactic nature is focused on the development of skills for effective operation in real life situations. This fact determines our choice of project-based learning as a tool for building axiological competence, promoting adaptation of future graduates to changing social and economic conditions and their successful performance in real professionally significant circumstances.

2. Governing equations

Considering the essentials for the formation of axiological competence, it is worth mentioning that competencies are the basis of competitive graduates. In fact a competence characterizes a process [1, p. 33], an area of activity [2], which a person is well aware of, has relevant knowledge and experience. A person competent in a specific field possesses appropriate knowledge and skills to enable him to judge reasonably about this field and to work effectively. Axiological competence, according to the classification of A.V. Khutorskoy [3] belongs to the key competences.

Axiological competence, as we understand, is characterized by the following: the ability to see, understand, cherish the natural world and nature; take scientific knowledge as a value; be able to perfectly adapt to the modern world, choose the system of values, targets and meanings for one's innovative activities, identify one's own contradictions and make decisions. Further livelihoods of people, their self-determination and self-development largely depend on this competence. Moreover, axiological component is included into all other competencies.

Like any other competencies axiological competence is based on activity component. From the structural perspective axiological competence integrates knowledge, environmentally oriented thinking as the quality of a person and a number of skills. The most important, in our opinion, is the ability to formulate one's own goals of innovation, comparing them with the values of the modern world; to make decisions under irregular circumstances, realizing the value and essence of one's activities; select and implement individual educational trajectory in accordance with personal goals.

Based on the foregoing assumptions, the system of building axiological competence of future graduate students, who study energy-saving technologies, include the following tasks:

1) formation of environmentally oriented thinking (new mentality associated with views on the environmentally sound and sustainable development of society);

2) development of independent thinking (the ability for self-development, self-sufficiency in getting new knowledge, skill to self-identify contradictions in dealing with complex educational and cognitive problems);

3) provision with methods of innovation.

Pedagogical conditions of building axiological competence of graduate students in learning energy-saving technologies are the following: development and application of special assignments directed at the formation of independent thinking through the solution of comprehensive educational and cognitive problems; the use of environmentally focused project-based learning; organization of training based on interactive and information technology.

Technique of axiological competence building includes a set of methods, tools, approaches and means used in the educational process in order to achieve a particular goal. We believe that project learning is an effective tool for the formation of the value attitude to scientific knowledge, for independent innovations with the view to harmonize the relations in the system "man-nature-society" [4]. The following part of the article will describe requirements for project-based learning and the specifics of its application.

Project culture is seen today as a mandatory component of the general culture of a modern man. [5] Students' project culture elements establishment was described in the papers of Russian teachers. However, the problems of personal development and the formation of axiological competence of future graduates through project-based learning of energy saving technologies have not still been resolved.

Project-based teaching method is aimed at the acquisition of new knowledge in close connection with the real-life practice. Project-based learning allows to engage students in active learning process by setting problems and choosing possible solutions, collecting necessary information, etc., thus acquiring new training and life experience.

Let's consider the example of selecting relevant topics for environmentally focused energy saving projects. It is commonly known that, one of the major challenges of the XXI century is a global environmental crisis caused by the anthropogenic impact on the environment. For example, according to research Steffen A., Douglas T., Amyot M. et al. [6], as a result of fuel combustion, mercury, contained in industrial emissions, is oxidized in the atmosphere. In the future, due to the metabolism of microorganisms oxidized mercury is transformed into highly toxic methyl mercury, which by its toxic effect outranks the metal itself. Based on this scientific fact one can formulate several project topics related to the study of processes of movement and control of mercury compounds in the environment, taking into account energy saving measures. Implementation of these projects necessitates integrated knowledge, creativity and environmentally oriented thinking. It is virtually impossible to select the most "reasonable" topics among the mentioned ones, as project learning is in general based on the creativity of teachers and students.

Choosing the project topics it is necessary to consider the possibility of 1) formulating an interesting problem (or problems), which require integrated knowledge and research approach to solve it; 2) theoretical, practical or cognitive significance of the expected results; 3) independent or group activities of the project participants; 4) structuring an informative part of the project (specifying step-by-step results); 5) application of research methods (hypothesis suggestion, theoretical justification of the problem solution hypothesis, choice of research methods, data analysis, its presentation, correction, conclusions drawing).

The specifics of energy-saving technologies project-based learning consists in the fact that it is expected to focus on the humanization of the course (historical and methodological, environmental and applied aspects).

During the implementation of the historical and methodological aspects of the project methodological issues are revealed. For example, a history of a substance as a part of nature; history of industrial production; history of development and establishment of energy saving; life and work of scientists in the field of energy saving and efficiency. These projects form the students' value attitude to integrated knowledge, broaden their horizons and establish interdisciplinary communication to reconstruct the complex evolution of scientific knowledge, to show the value and role of scientific prediction. The environmental aspect allows to expand the role of science in the fight against environmental problems; to attract students' attention to research on the state of the environment; foster a sense of personal responsibility for its preservation. Working on the project, students build the environmentally focused axiological competence, live and work without destroying the world, consciously participating in nature protection. Applied aspects of the project allow to give meaning to learning activities, to expand scientific and technical outlook. Applied research projects are easy to implement in such areas as energy and manufacturing; use of energy sources in industry; energy-saving technologies in heating, water, gas, and electricity supply, maintenance of microclimate.

The mentioned aspects may be integrated into one project. This is the specifics of energy-saving technologies project-based learning. The procedure of the project includes, as a rule, five stages: exploration, analysis, practice, presentation and control.

Axiological competence possesses various characteristics, which were discussed above. In this paper, we consider the possibility of building valuable attitude to scientific knowledge in the context of the main learning stages. This process is considered in the context of energy-saving technologies project-based learning for graduate students with a major in Biotechnology.

Graduate students have not received topics for group projects immediately. Initially, they were offered exemplary individual mini-projects, such as "Energy as a cultural phenomenon", "The most important regularities in the development of energy conservation and efficiency," "Development of state power supply ", "Problems of nuclear energy", "The problem of clean fuels", and others. Then graduate students performed group projects according to their choice from the following list: "Alternative and renewable energy in the energy economy of the state," "Power management and energy efficiency in the urban economy", "Energy saving management of industrial enterprises", "Energy supply and energy efficiency in agriculture" "Energy-saving technologies in crop and mobile energy", "Energy-saving technologies in animal husbandry and stationary power."

3.Conclusion

The effectiveness of using project-based learning for building axiological competence (the ability to accept natural sciences as a value) was determined by three main criteria (education, motivation and activity). The obtained results show the positive dynamics in the development of abilities to identify the problem and plan the work, search (research), communication, presentation and reflective skills that are so important for graduate students.

References

[1] Pisareva S.A. *Development of the concept of «competence» in the dissertation research on pedagogy. Academic reading. St. Petersburg: Herzen University* 2002 Vol. 3 P. 29–34

[2] Gilmanshina S.I. *Competence approach in higher professional education* 2007 Kazan: School

[3] Khutorskoi A. *Didactic heuristic. Theory and technology of creative learning* 2003 Moscow: MSU, 416 p.

[4] Suriya I. Gilmanshina, Rimma N. Sagitova et al. *Professional Thinking Formation Features of Prospective Natural Science Teachers Relying on the Competence-Based Approach. Review of European Studies*, Vol 7, No 3(2015) P. 341-349

[5] Sakhieva R.G., Gilmanshina S.I., Gilmanshin I.R., et al. *A Portfolio as an Alternative Means of Presenting the University Student's Achievements. Asian Social Science*; Vol. 11, No. 3; 2015. P. 162-167

[6] Steffen A., Douglas T., Amyot M. et al. *A synthesis of atmospheric mercury depletion event chemistry in the atmosphere and snow. Atmospheric Chemistry and Physics*, 2008 V. 8 P. 1445–1482

IOP Conf. Series: Materials Science and Engineering **86** (2015) 012030 doi:10.1088/1757-899X/86/1/012030

Mathematical model of the flow of polymer melt in the extruder forming die

R F Sharafutdinov[1], E R Galimov[1] and B A Snigerev[2]

[1]Kazan National Research Technical University named after A N Tupolev, Russia
[2]Institute of Mechanics and Engineering, Kazan science center, Russian Academy of sciences, Russia

E-mail snigerev@imm.knc.ru

Abstract. Steady creeping flow of viscoelastic fluid with the free surface which is realized when fluid polymer is entering the forming channel and leaving it is investigated. Fluid motion is described by equations conservation of mass and momentum supplemented by Metzner's determinative rheological constitutive equation of the fluid. On the basis of finite element method a stable numerical algorithm for solving the problem was developed.

1. Intoduction

The continuous growth of the production of chemical fibers and yarns requires the solution for a number of issues, among which should be noted such as improving the continuity of technological processes and their intensification, creation of a new high-efficiency equipment, reducing labor intensity by mechanization and automation of production processes. One of the main directions of development of mechanical engineering for the production of the finished products from polymer melts is an increase of spinning rate and combination of technological operations, decrease of energy and metal intensity of the equipment, mechanization and automation of processes using microprocessor technology. When solving all the above problems the development of mathematical models, methods of computation and optimization of molding process by extrusion with subsequent verification and implementation of the obtained results into production is becoming particularly urgent [1,2].

2. Governing equations

Motion of an incompressible viscoelastic fluid is described by the equations of conservation of mass and momentum.

In a cylindrical coordinate system creeping flow of a viscoelastic fluid in the absence of gravity is described by the system equations conservation of mass and momentum which are closed by the Giesekus determinative rheological constitutive equation [6, 7]:

$$\nabla \cdot \underline{v} = 0, \quad (1)$$

$$\rho\left(\frac{\partial \underline{v}}{\partial t}+\underline{v}\cdot\nabla\underline{v}\right)=-\nabla P+\nabla\cdot\underline{\underline{\tau}} \quad (2)$$

Here ρ – density of polymer melt, P – pressure, $\underline{v}$ – velocity vector, $\underline{\underline{\tau}}$ – stress deviator. The system of equations (1) - (2) is supplemented by the rheological constitutive relation connecting the stress deviator $\underline{\underline{\tau}}$ with the strain velocity tensor $\underline{\underline{D}}$ in the Metzner form

$$\underline{\underline{\tau}}=\underline{\underline{\tau_V}}+2\eta_N\underline{\underline{D}},\quad \underline{\underline{\tau_V}}+\lambda\overset{\nabla}{\underline{\underline{\tau}}}=2\eta_V\underline{\underline{D}} \quad (3)$$

where λ – fluid relaxation time, η_V, η_N – efficient dynamic viscosities of solvent and polymer melt, respectively. Dependence of viscosity on shear rate has the form

$$\eta_V=(\eta_0-\eta_\infty)\left[1+(k\,I_2)^2\right]^{\frac{n-1}{2}}+\eta_\infty,$$

here η_0, η_∞ – the greatest and the smallest effective shear viscosities of the polymer, I_2 – second invariant of the strain velocity tensor, k, n – model parameters. Operator $\overset{\nabla}{\underline{\underline{\tau}}}$ defines the upper convective derivative of a tensor as

$$\overset{\nabla}{\underline{\underline{\tau}}}=\frac{\partial\underline{\underline{\tau}}}{\partial t}+\underline{v}\cdot\nabla\underline{\underline{\tau}}-\underline{\underline{\tau}}\cdot\nabla\underline{v}-\nabla\underline{v}^T\cdot\underline{\underline{\tau}}$$

Polymer flow in the computational region is described by the equations (1) – (3) velocity profiles and complete stress distributions are specified at the entrance boundary of the computational region v = $v_0(x_2)$, $\underline{\underline{\tau}}_V=\underline{\underline{\tau}}_{0v}(x_2)$, steady uniform profile of velocity and stresses is specified at the region outlet $v=0$, $\partial u/\partial x=0$, $\partial\underline{\underline{\tau}}_V/\partial x=0$, kinematic and dynamic boundary conditions in the following form should be satisfied at the free boundary

$$\underline{v}\cdot\underline{n}=0,\quad \underline{n}\cdot\underline{\underline{\tau}}\cdot\underline{n}=2H/\chi,$$

where $\underline{n}$, $\underline{t}$ – unit normal vector and unit vector tangent to the free surface, $2H$ – radius of curvature of the free surface, χ – surface tension coefficient.

For the free surface described by the equation $F(\underline{x},t)=0$ the following relationship is satisfied

$$\frac{\partial F}{\partial t}+\underline{v}\cdot\nabla F=0 \quad (4)$$

Approximation of equations (1) – (3) and calculations are carried out by the finite element method of the second order on irregular meshes condensing to the zone of outflow of the polymer from the nozzle. Since the desired functions vary significantly only in the vicinity of outflow of the polymer, the use of fine meshes in this zone and sparse meshes outside it plays a significant role in saving computational costs. The sequence of condensing meshes of 9-node quadrilateral elements (the number of nodes 2000, 8400) was created for the calculations. Linear quadrilateral elements were used for the calculation of the stresses. Location of the deformable free surface is found from the approximation of the kinematic condition (4) by the finite element method then mesh of finite elements near it is rearranged to obtain solutions of matrix equations (1) – (3) using which the field of velocities, pressures and stresses at the new time step is found. Stationary solution of the problem is found by establishing the evolutionary problem using traditional algorithms for this class of equations.

3. Results

The problem of outflow of Newtonian fluid from the pipe at different Reynolds numbers was solved for approbation of the numerical algorithm. Increase of diameter of the output stream is characterized by parameter h_f which is equal to the ratio of stream diameter to the channel diameter and is called the coefficient of die swell. The figure shows the forms of the free surface of a Newtonian fluid at various Re numbers: 1 – 4.09; 2 – 12.5; 3 – 17.2; 4 – 27.3; 5 – 47.4 as compared to the experimental data [3] (solid symbols).

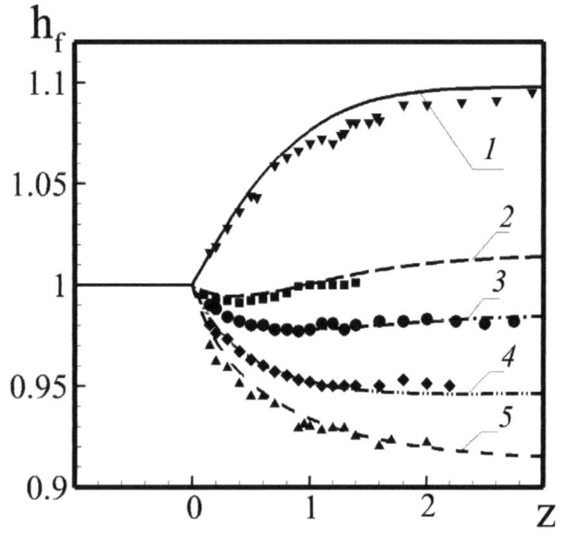

Figure. Forms of the free surface of Newtonian fluid at various Re: 1 – 4.09; 2 – 12.5; 3 – 17.2; 4 – 27.3; 5 – 47.4

It should be noted that there is satisfactory agreement between the calculated data and the experimental results which have an error of approximately ±0.01.

The obtained results show that effective finite element algorithm for calculating flows of viscoelastic fluids in regions with moving boundaries was created.

References

[1] Torner R V *Theoretical basics of polymer processing* 1978 M: Khimiya 426

[2] Yankov V I and Glot I O and Trufanova N M and Mitrofanova O V *Polymer flows in the die holes. Theory, calculations, practice* 2010 Publishing house URS Izhevsk 368

[3] Gear R L and Keentok M and Tanner R I 1987 The shape of Newtonian jet *Phys.Fluids.* **26** 7-11.

IOP Conf. Series: Materials Science and Engineering **86** (2015) 012031 doi:10.1088/1757-899X/86/1/012031

Investigation of influence of the outlet channel length of extruder die hydrodynamics of extrusion process

R F Sharafutdinov[1], E R Galimov[1] and B A Snigerev[2]

[1]Kazan National Research Technical University named after A N Tupolev, Russia
[2]Institute of Mechanics and Engineering, Kazan science center, Russian Academy of sciences, Russia

E-mail snigerev@imm.knc.ru

Abstract. Influence of the outlet channel length of an extruder on a steady creeping outflow of viscoelastic fluid is investigated. Fluid motion is described by equations of conservation of mass and momentum supplemented by Giesekus's determinative rheological constitutive equation of the medium. On the basis of finite element method a numerical algorithm for solving the problem was developed. Distribution pattern of fluid velocity in an outflowing jet, pressure and stresses is analyzed; results of dependency of swelling degree of polymer fluid on die geometry and parameters of a rheological model are investigated.

1. Intoduction

In the process of flowing of polymer melts and solutions through the forming channels of processing equipment different specific effects such as Barrus effect occur. This effect is of great importance in the process of product forming of polymer materials by extrusion method and represents an increase of outflowing jet diameter D_e extruded from the die compared to die diameter D_w. In technological practice it is estimated by a coefficient h_f which is equal to the ratio D_e/D_w [1-5]. Narrowing of outflowing jet of Newtonian fluid at high flow velocities is shown in work [1] and for low velocities (at Reynolds numbers less than 16) widening of outflowing jet is approximately 1.1 [2]. Application of numerical methods to obtain the flow structure and the form of outflowing jet is discussed in works [3, 4]. Severity of exposure of various factors on the extrusion character of polymer melt described by the Oldroyd rheological relation is discussed in work [5]. It is shown that the swelling during extrusion from flat channels exceeds similar values for axisymmetric channels for the same values of average velocities of the polymer at the inlet.

2. Governing equations

This work considers outflowing of a viscoelastic fluid from cylindrical stepped channel its scheme is presented in figure 1. In the figure h denotes a radius of the narrow channel of the jet outlet part, $4h$ – radius at the tube inlet, $h_1 = 0.5h$ – size of the rounded part, L – length of the die outlet channel.For calculations values of L are chosen to equal to $L = 0, 2h, 4h, 10h$. In this case we analyze the influence of outlet channel length on the structure of the output flow of the polymer fluid in the extrusion process as well as investigate the influence of rheological parameters of the Giesekus constitutive relation characterizing the level of viscoelastic properties of the polymer material.

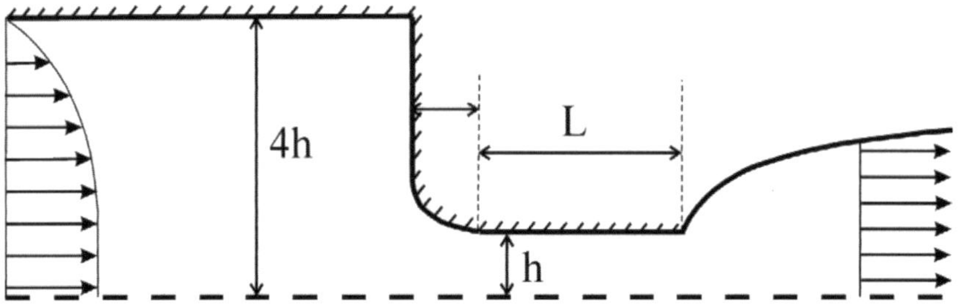

Figure 1. Scheme of outflowing of the viscoelastic fluid from the cylindrical stepped channel.

In a cylindrical coordinate system creeping flow of a viscoelastic fluid in the absence of gravity is described by the system equations conservation of mass and momentum which are closed by the Giesekus determinative rheological constitutive equation [6, 7]:

$$\rho\left(\frac{\partial u_i}{\partial t}+u_j\frac{\partial u_i}{\partial x_j}\right)=\frac{\partial \sigma_{ij}}{\partial x_j},\qquad \sigma_{ij}=-P\delta_{ij}+\tau_{ij},\quad \frac{\partial v_i}{\partial x_i}=0, \tag{1}$$

$$\tau_{ij}=\tau_{ij}^V+2\eta_N D_{ij},\quad \frac{\delta\tau_{ij}^V}{\delta t}=\frac{\partial\tau_{ij}^V}{\partial t}+u^k\frac{\partial\tau_{ij}^V}{\partial x^k}-\frac{\partial u_k}{\partial x_i}\tau_{kj}^V-\tau_{ik}^V\frac{\partial u_k}{\partial x_j}, \tag{2}$$

$$\tau_{ij}^V+\lambda\frac{\delta\tau_{ij}^V}{\delta t}+\frac{\chi_G\lambda}{\eta_V}\tau_{ij}^{V\,2}=2\eta_V D_{ij},\quad D_{ij}=\frac{1}{2}\left(\frac{\partial u_i}{\partial x_j}+\frac{\partial u_j}{\partial x_i}\right) \tag{3}$$

In the system of equations (1) – (3) u_i – fluid velocity components, P – pressure, σ_{ij} – total stress, τ_{ij} – stress deviator, τ_{ij}^V – viscoelastic part of the stress, δ_{ij} – unit tensor, ρ – density, η_V, η_V – dynamic viscosities of polymer and solvent respectively, λ – stress relaxation time of a polymer fluid, χ_G – dimensionless parameter of Giesekus rheological model.

For a free surface described by the equation $F(x_i,t)=0$ kinematic relation is satisfied [8]

$$\frac{\partial F}{\partial t}+u_i\nabla F=0 \tag{4}$$

Approximation of the equations (1) - (4) and calculations are carried out by finite element method of second order with linear and quadratic approximations of the unknown functions [2].

3. Results

Steady extrusion of polymer fluid from the stepped cylindrical die with the following physical properties: ρ = 924 kg/m^3; $\eta_V = 1.2\cdot10^4$ Pa·s, $\eta_N = 0.9\cdot10^3$ Pa·s – shear viscosity at zero shear rate is considered. Channel radius of a narrow part of the die $h = 10^{-2}$ m, average velocity of a supplied fluid at the inlet $U_0 = 4.0\cdot10^{-3}$ m/s. Relaxation time characterizing the viscoelastic properties of a fluid was varied in the range of λ =0.1-10 seconds. For such values of selected parameters the following values of dimensionless criteria are obtained: Reynolds number determined as $\mathrm{Re}=\rho U_0 L/(\eta_V+\eta_N)=10^{-4}$, Weissenberg number $We = \lambda U_0/L$, varying in the range of 0.1-5. As U_0 an average fluid velocity at the inlet is considered, L – characteristic dimension, which is equal to the radius of a narrow part of the channel h.

Paper [5] presents experimental data on the degree of swelling of a viscoelastic fluid outflowing from a cylindrical channel in case when polystyrene was used as a polymer fluid. For comparison with experimental data fluid Oldroyd-B is selected as rheological model its constitutive equation is described by equations (1) – (3). Figure 2 shows comparison of calculated and experimental data h_f [5] in the range of *We* numbers from 0.1 to 5.

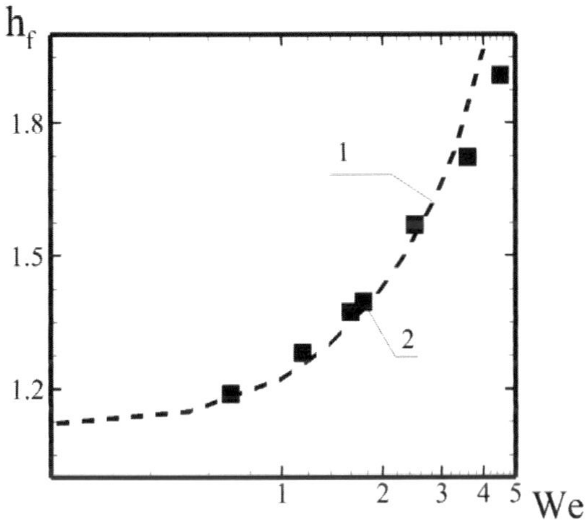

Figure 2. Dependency of h_f on *We* for the fluid Oldroyd-B: curve 1 – calculations, curve 2 – experiment [5].

Quite satisfactory agreement of the numerical predictions for different grids with experimental data indicates the adequacy of the developed numerical methods for calculating flows of Newtonian and viscoelastic fluids with free surfaces. Figure 3*a* shows a dependency of swelling degree h_f of outflowing jet of a polymer fluid on the value of *We* number for different dies. Forms of the dies differ with length of narrow outlet part *L* varying from 0 to 10*h*. Results show that swelling degree h_f of the jet increases with decreasing of the length of outlet section. It can be seen that the values h_f depend on the kinematics of the preceding deformation - shear rate and length of the output part of the capillary. This can be explained by the fact that post-extrusion swelling in the processing of polymers is caused by the release of elastic energy stored during the previous flow in a channel. Viscoelastic fluids are the mediums possessing a "hereditary" memory. A hereditary fluid remembers prehistory, i.e. the change of the velocity field in the previous times. Duration of memory is characterized by time of relaxation processes λ, which is determined by the parameter *We*. It can be seen that with increasing the *We* number the swelling degree increases for all dies. The investigation results of influence of a rheological parameter of the Giesekus model χ_G on a swelling process under an isothermal flow is

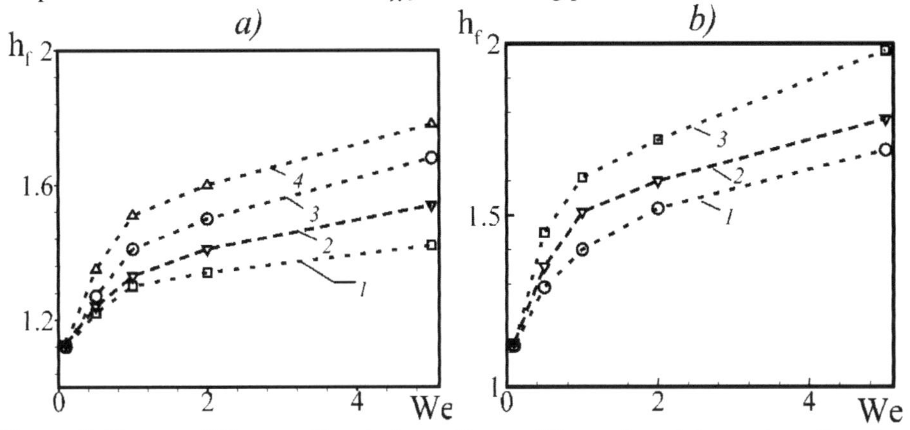

Figure 3. Degree of Barrus effect h_f depending on *We* number *a*) for various dies: 1 – L = 10h, 2 – L = 4h, 3 – L = 2h, 4 – L = 0;
b) for various values of parameter χ_G for the die with length $L = 0$, 1 – $\chi_G = 0.86$, 2 – $\chi_G = 0.33$, 3 – $\chi_G = 0.1$.

shown in figure 3*b*, which shows the dependency of a swelling degree of the outflowing jet of a polymer fluid on the value of *We* number for the die with the length $L = 0$ when increasing the parameter χ_G of the Giesekus model. The increase of this parameter for viscoelastic fluids means that they possess a greater viscosity anomaly with increasing the shear rate. Fluids which does not possess a viscosity anomaly are described by Oldroyd rheological equation [5] this corresponds to $\chi_G = 0$ in Giesekus model. It can be seen that for viscoelastic fluids showing great viscosity anomaly the swelling degree decreases during extruding from the die. This is explained by the fact that the elastic properties of a fluid play a significant role in increasing the diameter of the output jet. For low values of stress relaxation time (respectively *We*) behavior of the polymer mixture is close to a Newtonian fluid when the value h_f does not exceed 1.18. Figure 4 shows, as an example, a flow pattern for the die with the size of the output channel equal to $L = 2h$ in the form of the streamlines (figure 4*a*) and contours of the first difference of principal stresses (figure 4*b*) at $We = 3.0$. From the presented data it can be seen that as soon as the polymer fluid passes the outlet section and enters the zone of a free flow pressure and stresses immediately begin to relax.

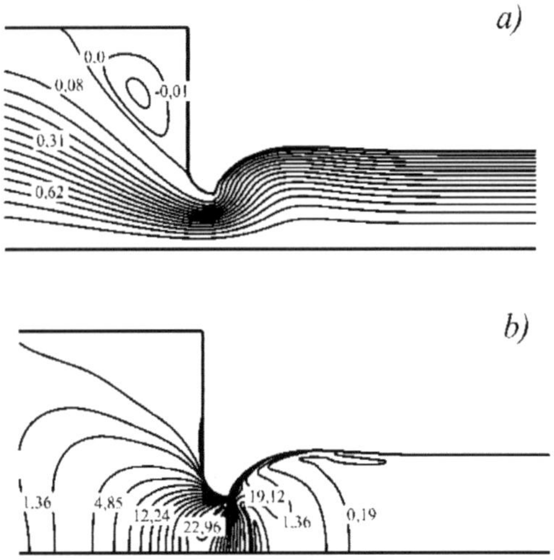

Figure 4. Streamlines ψ (*a*); contours $N_1(b)$ when $We = 3.0$ for the die $L = 0$.

The relaxation process takes place until the accumulated highly elastic deformation falls below a value corresponding to the elastic deformation of the polymer in a state of steady flow. We distinguish two main reasons for the change of the jet cross-section:

- reconstruction of a velocity profile from a parabolic corresponding to the flow in a channel, to a rectangular corresponding to the jet movement as a continuous solid rod;
- relaxation of the highly elastic tensile strains accumulated in the jet material during its passage through the channel.

Figure 5 presents fluid velocity profiles for the five cross-sections for the die of length $L = 0$. These data show the influence of the viscoelastic properties of the fluid on the character of their outflow from the die. For low values of *We* number the behavior of polymer fluids due to a slight appearance of viscoelastic properties is close to Newtonian (figure 5*a*). The figure also shows that redistribution of the velocity profile from parabolic to one-dimensional occurs at the channel outlet with increasing the distance. In this case influence of viscoelastic properties of the fluid on the swelling degree and change of the character of velocity redistribution in the outlet flow becomes apparent. The general trend is reduction of the velocity along the central axis with increasing distance and the difference is that for viscoelastic fluids, because of the greater swelling the steady speed in the output flow is significantly less. One of the essential nonlinear distinctions is the difference in the velocity distribution along the surface of the output flow. If for a Newtonian fluid a velocity along the surface increases monotonically up to a steady value, for viscoelastic fluid velocity on the surface increases and then decreases.

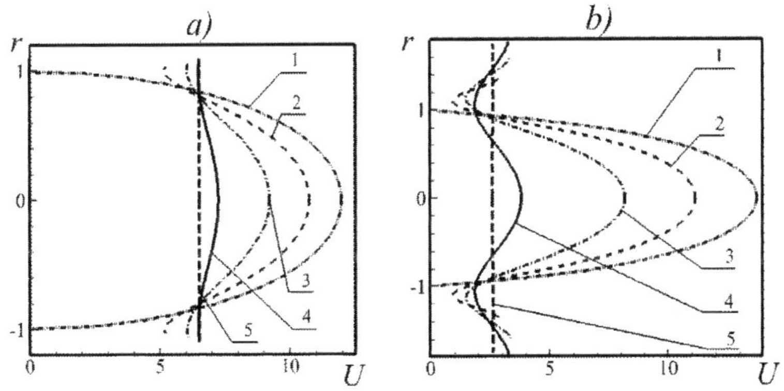

Figure 5. Velocity profiles in transverse cross-section in the output jet for 1 – z/h = 0.0, 2 – 0.50, 3 – 0.75, 4 – 1.0, 5 – 3.0 for the die $L = 0$ when $We = 0.1$ (*a*) and $We = 3.0$ (*b*).

Figure 5*b* shows that steady-state value of the velocity at the border for the cross section z/h = 3.0 (curve – 5) is less than the values for z/h = 0.75, 1.0 (curves 3 and 4). Overspeeding at the border is not so significant compared with the Newtonian case. More significant difference appears in the character of the velocity distribution across the flow. For a Newtonian fluid minimum velocity value is at the jet border which then monotonously increases toward the centre of the flow. Influence of viscoelastic fluid properties reflected in the fact that the velocity in cross-section is minimal not on the flow border, but is located on a certain deepening from the surface. Velocity profile across the flow is nonmonotonic, the maximum velocity at the central axis of the flow decreases as it approaches the border, but then near the surface it begins to increase again. Thus the greatest gradients of the velocity change are observed in the immediate vicinity of the channel outlet, and the minimum value of the velocity in the cross section begins to increase with increasing the distance from the outlet. Also as a significant difference it should be noted that a nonmonotonicity of the velocity in cross section is observed along the entire length of velocity establishment and reconstruction up to a homogeneous flow. The results of numerical modeling of the outflowing of a viscoelastic fluid from the die based on the Giesekus fluid model showed that for simulating the effect of fluid swelling it is necessary to choose a non-linear viscoelastic models that take into account the first and second difference of normal stress in shear flow. The existence of normal stresses is one of the reasons for the display of Barrus effect.

Presented results of the calculations of the extrusion process of a viscoelastic fluid from the stepped forming dies with different length of the output part showed that the swelling degree of the output jet increases for the short dies. These results support the conclusion that one of the main factors influencing the swelling degree of the hereditary viscoelastic mediums is a stress relaxation time.

References

[1] Nickell R E and Tanner R I and Caswell B 1974 The solution of viscous incompressible jet and free surface flows using finite element method *J. Fluid. Mech.* **65** 189-206

[2] Crochet M J and Keunings R 1982 Finite element analysis of die swell of highly elastic fluid *J. Non-Newtonian Fluid. Mech.* **10** 339-356

[3] Russo G and Phillips T N 2011 Spectral element predictions of die swell for Oldroyd fluids *Comput. Fluids* **43** 107-118

[4] Mitsoulis E 2010 Extrudate swell of Boger fluids *J. Non-Newtonian Fluid. Mech.* **165** 812-824

[5] Brasseur E and Gergiou G C and Crochet M J The time dependent extrudate swell problem of a fluid with slip along the wall *Rheol. Acta.* **42** 549-566

[6] Giesekus H A simple constitutive equation for polymer fluids based on the concept of

deformation dependent tensorial mobility *J. Non-Newtonian Fluid. Mech.* 1982 **11** 69-109
[7] Bird R B and Armstrong R C and Hassager O 1987 *Dynamics of Polymeric Liquids. vol.1. Fluid Mechanics. 2nd ed.* New York John Wiley and Sons
[8] Silliman W J Scriven L E 1980 Separating flow near a static contact line: slip at a wall and shape of a free surface *J.Comput. Phys.* **34** 287-313

IOP Conf. Series: Materials Science and Engineering **86** (2015) 012032 doi:10.1088/1757-899X/86/1/012032

Prediction of samples failure during severe plastic deformation by multiple extrusion

V G Shibakov, D L Pankratov, A P Andreev, S I Andreeva

Kazan Federal University, Naberezhnye Chelny Institute
423810, Naberezhnye Chelny, Mira av., 68/19, Russia.

E-mail: andreevanton@list.ru

Abstract. A method of obtaining a hardened semi-intensive plastic deformation by extrusion is considered. The results of simulation process at different stages of extrusion are given. Conclusions are drawn about the applicability of the simulation results to obtain the axisymmetric products. The results of sample study are given according to the method of determining the criterion of damage Ψ .

1. Introduction
Significant refinement of metals and alloys structure until the occurrence of nanostructured states can be provided by severe plastic deformation (SPD).

Special schemes of mechanical deformation are known and studied, such as intensive torsion under high pressure torsion disc blanks on the Bridgman anvils, equal channel angular pressing (ECAP), comprehensive forging etc. Recently, a number of new technological SPD schemes have appeared: "hourglass" [1], screw extrusion, in which intense shear deformation is achieved by hydromechanical pressing prismatic billet through a matrix with a screw channel, which cross section is orthogonal to the axis of extrusion and continuously along this axis [2].

The above mentioned technological scheme of SPD has a number of advantages. First, it is possible to make massive samples for mechanical testing. Secondly, residual porosity, obtained from powder materials, can be substantially reduced. Thirdly, these methods can be used to obtain ultra fine-textured patterns not only in modeling, but also in industrial alloys that can lead to the prospects of industrial applications. A promising method of SPD metals processing by expression for grinding structures can be rated among these methods [3].

2. The Main Part
The basic technological stages of this method are shown in figure 1.

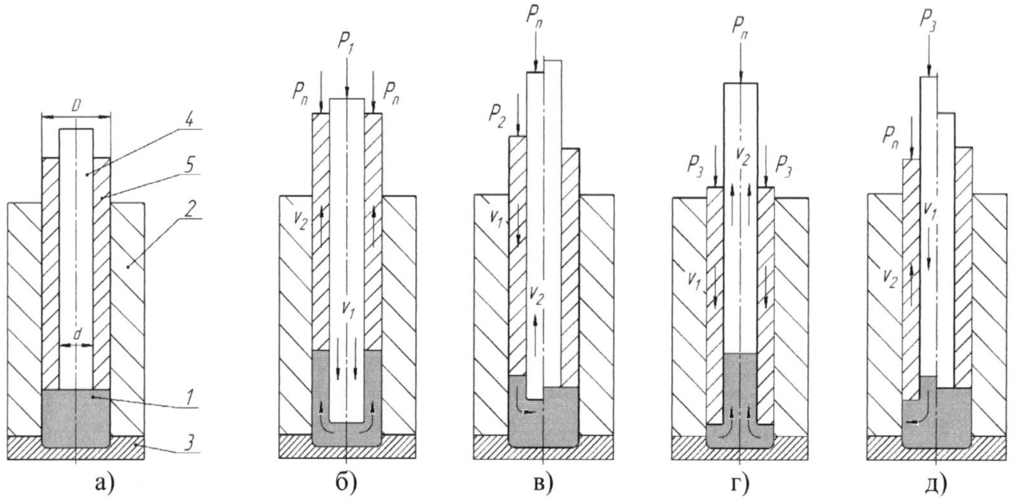

Figure 1. The process diagram of SPD by expression:
(a) initial state; b) backward extrusion of "glass"; b) formation of the initial shape of the radial extrusion of "glass"; g) backward rod extrusion with thickening; d) formation of the initial shape of the radial extrusion of the rod with thickening; 1 – blank; 2 – matrix; 3 – plate washer; 4 – solid punch; 5 – hollow punch; $P_1 - P_3$ – extrusion power at different stages; $P_п$ – back-pressure power ($0 \le P_п < P_2$, $0 \le P_п < P_3$); D, d – diameters of the punches; v_1, v_2 – tools speed

The workpiece 1 is placed in the matrix 2, mounted on a special plate 3. The height of the matrix exceeds the height of the workpiece. A solid 4 and a hollow 5 punches are put on the workpiece, which alternately carry out the deformation.

Figure 2 presents the algorithm, showing various combinations of deformation in SPD expression.

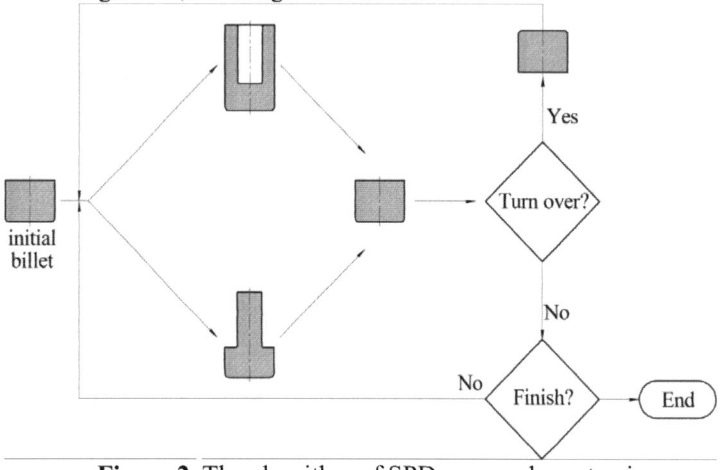

Figure 2. The algorithm of SPD process by extrusion

Based on the developed algorithm the matrix of different variants was formed (table 1) of SPD process by extrusion, where 0 and 1 are Boolean variables corresponding to the execution of deformation of 1, or not performing to 0.

The process may begin with a backward extrusion of "glass" (deformation by tool 4) or rod with a thickening (deformation by tool 5). Then the deformation is carried out by tool 5 or by tool 4, up to obtaining the initial contour of workpiece (for example, options 1 and 2 in table 1).

Table 1. Morphological matrix of SPD variants by extrusion SDI

Variant #	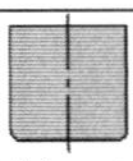 Initial contour of workpiece	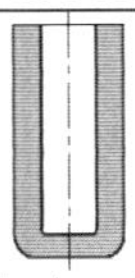 Backward extrusion of "glass"	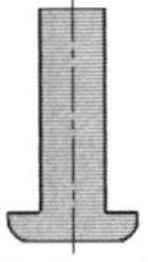 Backward extrusion of a rod with a thickening	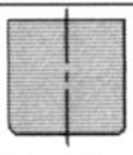 Radial extrusion with the receipt of the original form	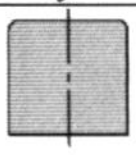 Casting-off	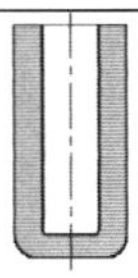	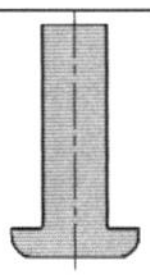	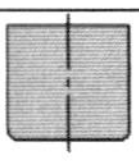
1	1	1	0	1	0	0	0	0
2	1	0	1	1	0	0	0	0
3	1	1	0	1	0	0	1	1
4	1	0	1	1	0	1	0	1
5	1	1	0	1	1	1	0	1
6	1	0	1	1	1	0	1	1
7	1	1	0	1	1	0	1	1
8	1	0	1	1	1	1	0	1

Structure formation and strain hardening during the transition to intensive plastic deformation, were first observed and systematically studied in the works of V.I. Trefilov, Y.I. Milman and S.A. Firstov [4-8]. With the growth of strain in the material a qualitatively new type of dislocation structure is formed that are strongly misoriented cells. Their main feature is the continuous increase of misorientation angles with strain increasing at a weak reduction in the transverse dimensions. This pattern evolution is inherent to other metals. V.I. Trefilov with employees suggested to evaluate strain hardening during intense deformation from the standpoint of the reduction of grain size in the extreme case up to size of strongly misoriented cells. His point of view about the radical changes in the mechanism of hardening received numerous confirmations and is now recognized by many scientists. Directed and continuously occurred motion of disclinations in the process of plastic deformation should lead to the fragmentation patterns of the metal, i.e. breaking it up into microregions, on misoriented angles of the order of several degrees. The more the degree of plastic deformation, the smaller should be the fragments and more is the turn relative to each other [9].

3. Experimental part

To establish the effect of the method on the stress-strain state (SSS) and determine the optimal deformation variants lead to the investigation of SPD process by extrusion in the QForm 7.2 environment. The simulation was performed sequentially according to the variant matrix (table 1). Material of the workpiece is alloy AD1, the temperature of the workpiece is 20°C, the temperature of the tool is 20°C, graphite grease. Strain rate is 1 mm/s. For easier comparison of the obtained results the scale of plastic deformation is he same in different extrusion variants and are graded from 0 to 20. The simulation results at different extrusion variants are summarized in table 2. The table is formed based on the algorithm shown in figure 3.

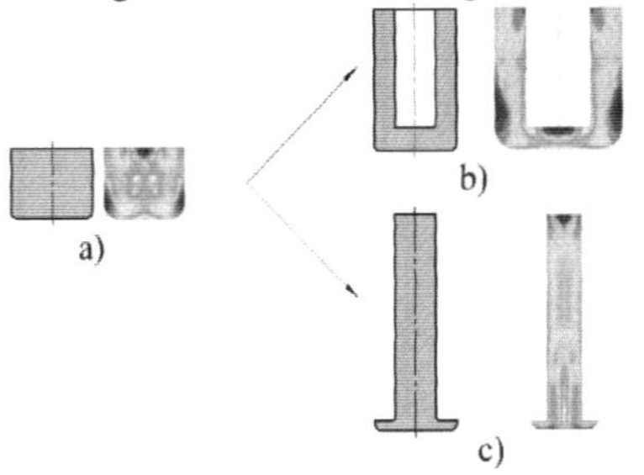

Figure 3. The algorithm of summary table formation of results (table 2): (a) workpiece with accumulated deformation rate after the implementation of one embodiment of the extrusion; (b) semi-finished type of "glass"; (c) semi-finished type of "stud with thickening"

Table 2. Table of simulation results for the various extrusion variants

	Variant 1	Variant 2	Variant 3	Variant 4	Variant 5	Variant 6	Variant 7	Variant 8
a)	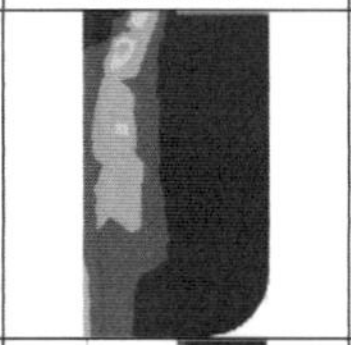	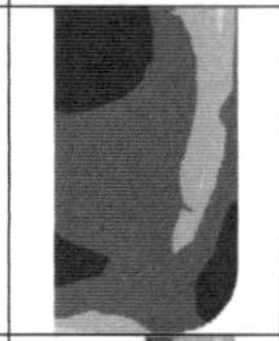	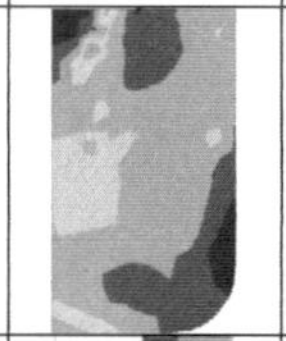	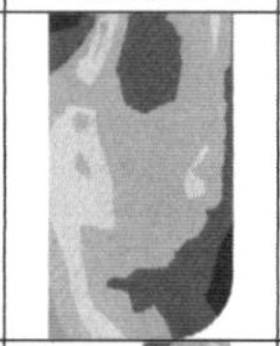	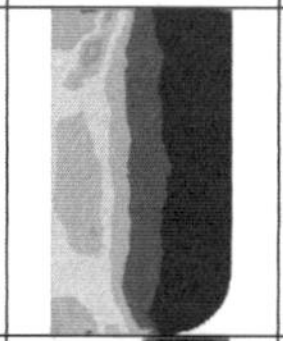	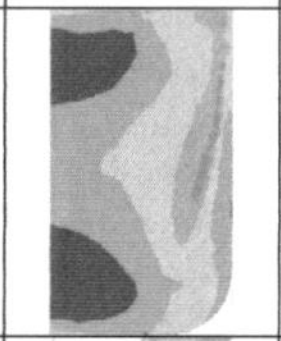	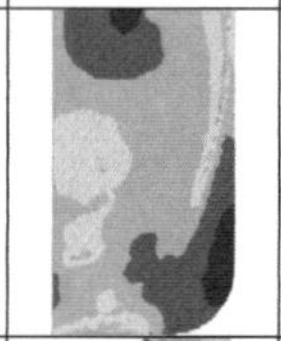	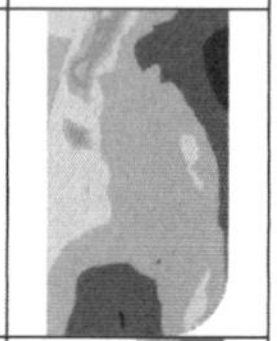
b)	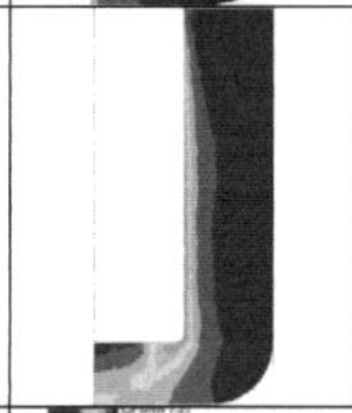	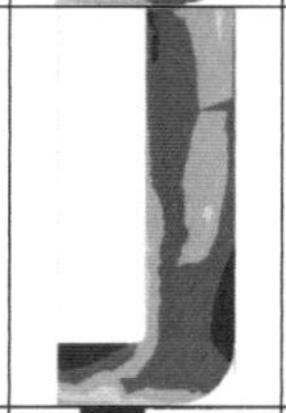	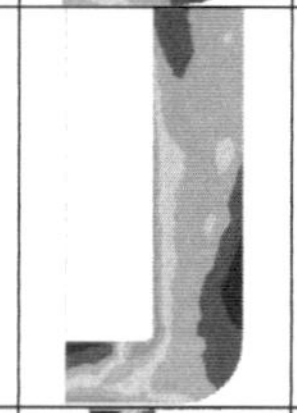	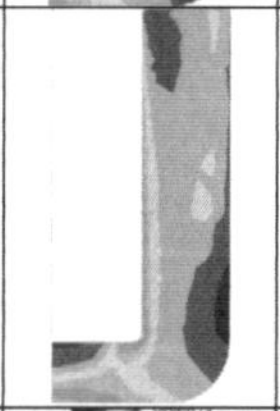	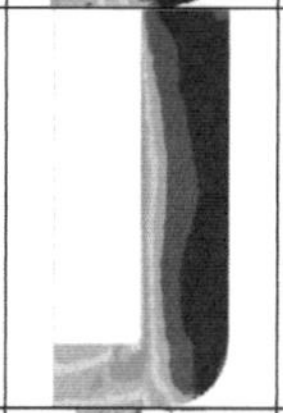	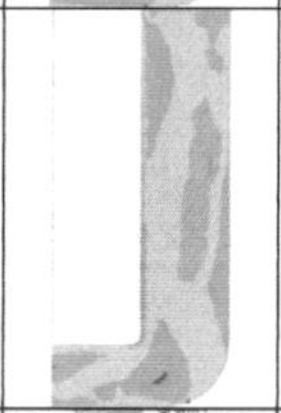	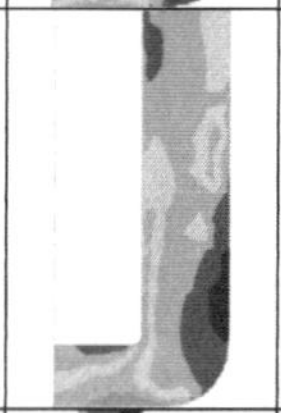	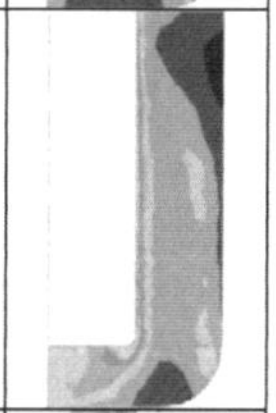
c)	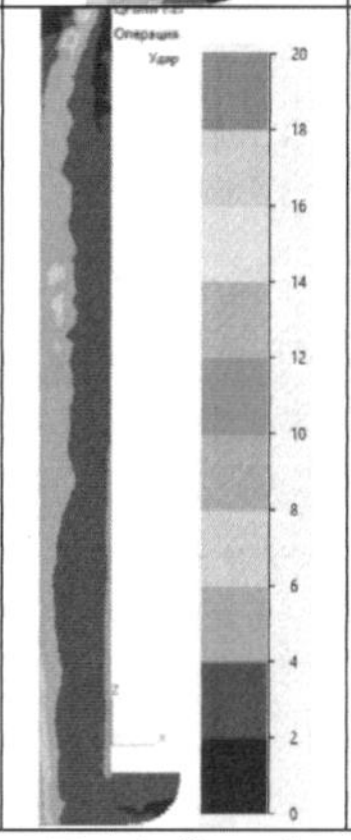	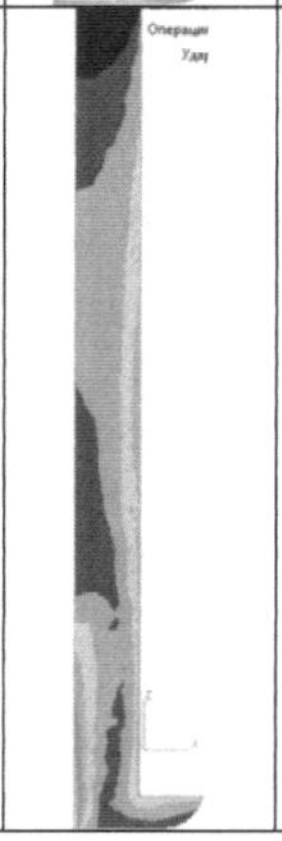	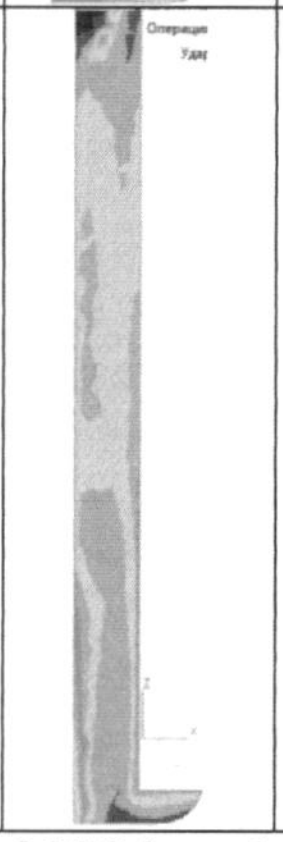	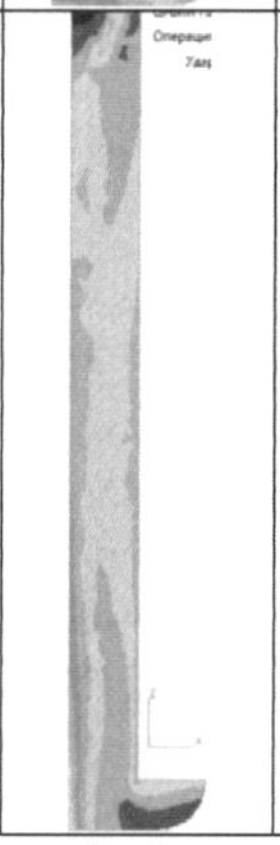	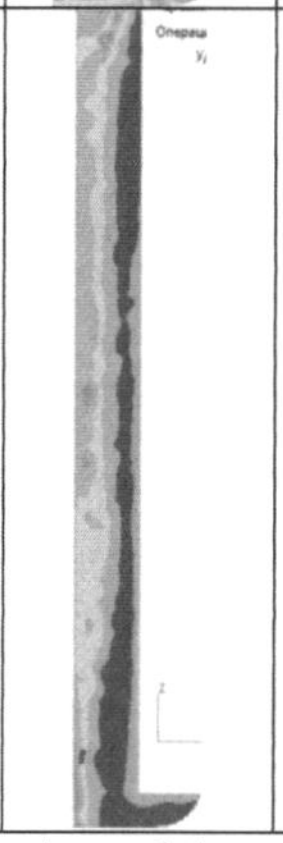	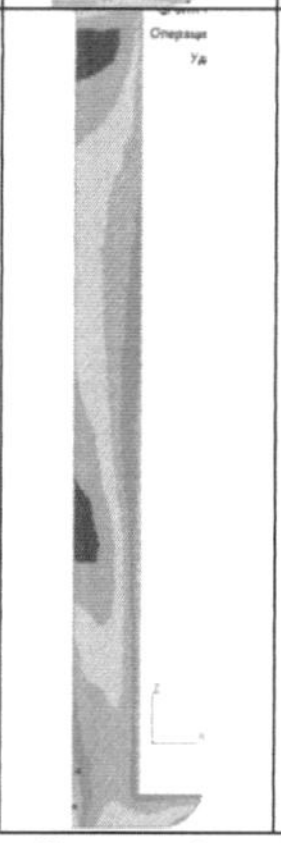	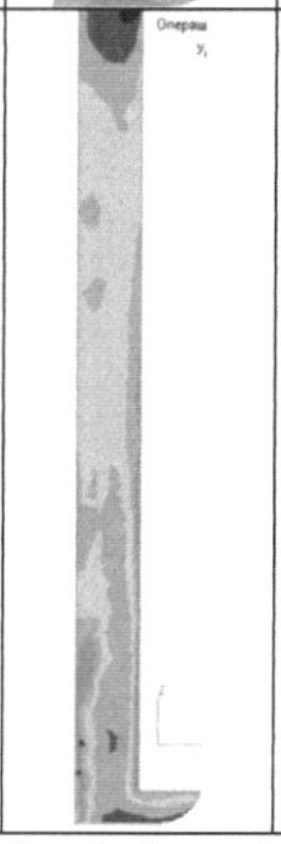	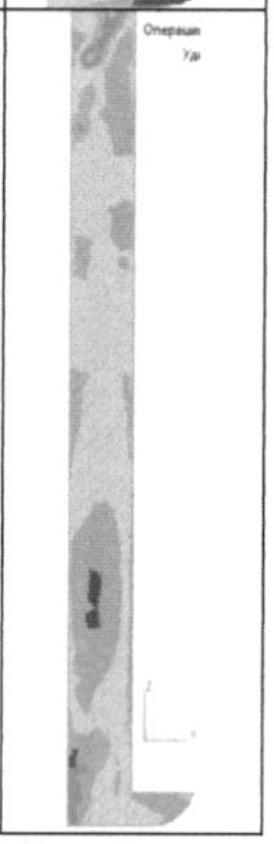

To define the parameters of SSS based on the interaction of the software for simulation of forming processes, and software that allows you to perform mathematical calculations in Excel and MathCAD environments, in areas where destruction is likely to occur, the history of deformation is traced. Next, this information in the form of numerical tables is passed to the software for mathematical calculations, where method of determining the criterion of damage Ψ is implemented. Further, there will be a check on the well-known V. L. Kolmogorov condition

$$\psi = \int_0^{E_i} \frac{d\varepsilon_i}{\varepsilon_{np}(\eta)} < 1 \qquad (1)$$

After a preliminary simulation of extrusion options of variants 1 and 2 (table 1), several areas were identified in workpieces where the destruction is possible. The traced points are placed in these zones (figure 4). The first zone includes points 2, 4, 5, 6, 7, 8, 9 and 10, which are located in the peripheral area of the workpiece, i.e., those points that are in the process of deformation directly interact with the tool and the matrix. The second zone includes points 1, 3, 11, 12, 13, 14, 15 and 16 – points inside the tested workpiece.

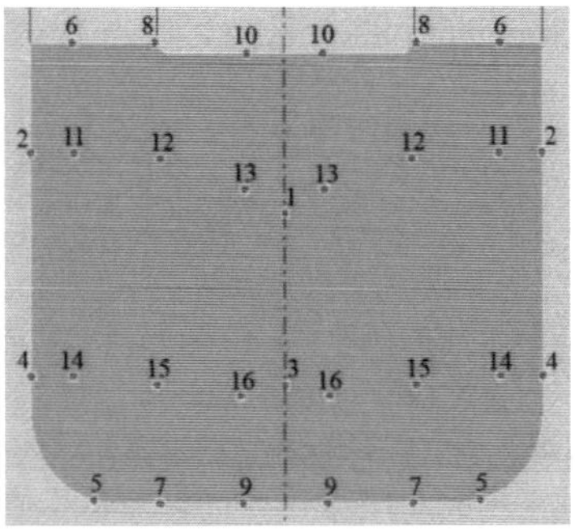

Figure 4. The location of the traced points after extrusion for option No. 1

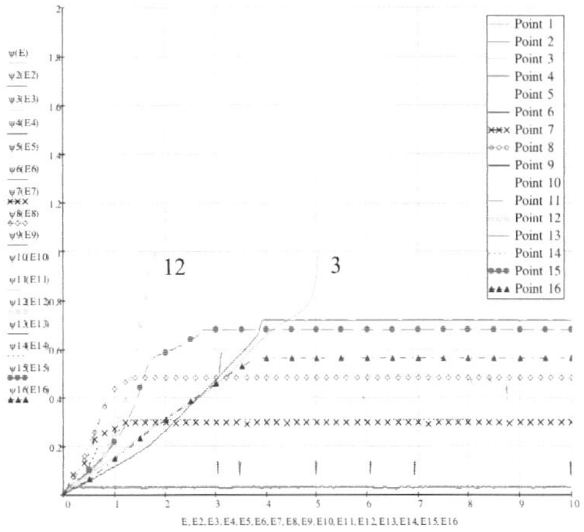

Figure 5. Changing of the used plasticity resource in the course of deformation for option No. 1

In points No. 12 and No. 3 (figure 5) plasticity resource is exhausted and we can expect the formation of defects. In general, after extrusion for option No. 1, an adequate supply of plasticity is present to continue the accumulation of deformation with the purpose of increasing the strength of the workpiece.

4. Conclusions

The discussed method is promising for the implementation of SPD in large-sized workpieces, including intermetallic compounds, and other materials.

SSS allows to provide intensive plastic deformation of workpieces without damage and defects.

The simulation results and SSS in the workpieces with the possible application of these results to obtain blanks or semi-finished products are compared.

The work was done due to the state support of Kazan (Volga region) Federal University in order to increase its competitiveness among the world's leading scientific and educational centers

References

[1] Golubev O.V. Development of high durability cold heading tool work piece production technology. Diss....Cand.Tech.Sci. Ufa, 1999. 146 p.

[2] Orlov D.V. Metal structure formation with a screw press//7th All-Russia scientific conference of physicist students: Thesis of papers. St.-Petersburg, 2001. http://asf.ur.ru/VNKSF/Tezis/v7

The patent №2189883 RF, ICI 7 B 21 J 5/00, 13/02, C 21 D 7/02.

[3] Method of plastic metal structurization at intensive plastic deformation and the device for its realization.

[4] Trefilov V.I., Milman J.V., Firstov S.A. Physical principles of strength of refractory metals. - Kiev: Naukova dumka, 1975. - 315p.

[5] Structure, texture and mechanical properties of wrought molybdenum alloys / V.I.Trefilov, J.V. Milman, R.K.Ivashchenko and others - Kiev: Naukova dumka. 1983. - 232 p.

[6] Rubtsov A.S., Rybin V.V. Structural features of plastic deformation at a stage of a flow localization//Physics of metals and metal science-1977 T. 44. Vol.3. - p.611-621

[7] Manilov V.A., Tkachenko V.T., Trefilov V.K, Firstov S.A. Structural changes in chromium at deformation//Metals – 1967-№2. - p.114-122.

[8] Zaripov R G, Tkachenko L A and Shaydullin L R 2014 *J. Phys.: Conf. Series* **567** 012042

[9] Rybin V.V. Big plastic deformations and destruction of metals. - M.:Metallurgy, 1986.-226 p.

Mathematical modeling and calculation of heating and melting of particles of the polymeric powder in flow channel of the sprayer

V L Fedyaev[1,2], I V Morenko[1], A R Siraev[1], E R Galimov[2], I R Gimranov[3], L R Fazlyev[3] and M S Takhaviyev[3]

[1] Institute of Mechanics and Engineering, Kazan Science Center, Russian Academy of Sciences, Russia
[2] Kazan National Resear ch Technical University named after A N Tupolev, Russia
[3] LLC "GazpromTransgaz Kazan", Russia

E-mail: morenko@imm.knc.ru

Abstract. Heating and melting of particles of polymeric powder in the central flow channel of spraying gun is investigated. Mathematical models of these processes taking into account convective and radiative-convective heat interaction of particles with environment is represented. Relations for calculating the temperature of the particles depending on the longitudinal coordinate, time of flight, operating and design parameters as well as thermophysical characteristics of the particles material and environment are given.

1. Introduction

When applying polymeric powdered materials (compositions) for retaining powder particles on the treated surface it is preheated to a temperature that ensures melting of the particles at the point of contact with the surface. Particles charging, their deposition on the electrode with subsequent melting and formation of the coating is also produced. In the case of flame spraying the powder particles are fed in the high temperature jet of the fuel combustion products. Thus the probability of material destruction on the particle surface and as a result appearance of the coating imperfections is high [1, 2]. In order to prevent the mentioned undesirable circumstances, taking into account the high thermal inertia of the material of sputtering polymeric powders it is appropriate to carry out heating and melting of the particles stage-by-stage: first in the proportioning feeder then in the first section of the central flow channel of the sprayer by convection through the dividing wall. After that in the second section in the area of direct contact of powder-air mixture with the gases heated up to high temperature the melting of the powder particles should be carried out mainly through the radiant heat exchange.

First section of flow channel of the sprayer consists of a central annular channel of length l_1 with a helical insert wherein the powder-air mixture is moving and a peripheral annular channel of the same length with a hot gaseous medium flowing in it. In fact, this section represents a recuperative tube-in-tube heat exchanger. Second section is a tube of length l_2 its inner diameter coincides with the diameter of the peripheral annular channel of the first section. Outer surface of this tube like outer surface of the tube of first section is thermally insulated. In the tube of the second section powder-air mixture and heated gaseous medium (air, fuel combustion products) covering the mixture flow are moving at different speeds.

When considering nonisothermal flow of mediums in the cavities of these sections it is assumed that mediums are incompressible and in the first section powder-air mixture is homogeneous. Cylindrical coordinate system r, φ, z is introduced; its axis $0z$ is directed along a symmetry axis of the concentric tubes.

2. First section of flow channel

Initial system of equations describing the motion of powder-air mixture in the central annular channel consists of differential equations of conservation of momentum, mass, energy and dependency of powder-air mixture density on the temperature.

At the first section equations of this system are averaged and supplemented by closing phenomenological relations.

In order to find formulas suitable for engineering calculations, the system of averaged equations is simplified as much as possible using the following assumptions. Turbulent nonisothermal flow of the mixture in the channel is considered quasi-stationary, terms containing z coordinate derivatives in the equations of conservation of momentum and continuity are neglected, change of the heat flow density in the longitudinal direction is assumed to be small compared to the change in the radial direction [3].

Thus obtained problem is solved with regard to the weight-average temperature of the mixture

$$\bar{t}_1 = \bar{t}_1(z) = \bar{t}_{01} + 2\pi r_1 q_c z \big/ \left(\rho_1 c_1 S_1 \bar{u}_1\right), \tag{1}$$

where $\bar{t}_{01}$ is the average temperature of the powder-air mixture at the inlet to the annular channel, $\bar{u}_1$ is the average speed of the mixture flow rate taken relative to the total cross section of the channel $S_1 = \pi\left(r_1^2 - r_0^2\right)$; r_0, r_1 is the radius of the central cylindrical insert; r_1 is the radius of the internal surface of the annular channel; ρ_1 is the average mixture density, c_1 is the average specific heat capacity of the mixture; q_c is the heat flow density from the side of heated gas.

Similar relationship can be found for the weight-average temperature $\bar{t}_2 = \bar{t}_2(z)$ of the heated gas in the peripheral annular channel. After a number of transformations of these relationships we get

$$\bar{t}_1 = \left(\bar{t}_{01} + k_l \bar{S}_1 \bar{t}_2\right) \big/ \chi_1, \tag{2}$$

$$\bar{t}_2 = \left(\bar{t}_{02} + \left(k_l \bar{S}_2 \big/ \chi_1\right) \bar{t}_{01} + \bar{S}_E q_E\right) \Big/ \chi_2 . \tag{3}$$

Here $\chi_1 = 1 + k_l \bar{S}_1$, $\chi_2 = 1 + k_l \bar{S}_2 / \chi_1$, $\bar{S}_1 = \bar{S}_1(z) = \pi z \big/ \left(\rho_1 c_1 S_1 \bar{u}_1\right)$, $\bar{S}_2 = \bar{S}_2(z) = \pi z \big/ \left(\rho_2 c_2 S_2 \bar{u}_2\right)$, $S_2 = \pi\left(r_2^2 - \bar{r}_1^2\right)$ is the area of a total cross section of the peripheral channel, $\bar{S}_E = \bar{S}_E(z) = S_E \big/ \left(\rho_2 c_2 S_2 \bar{u}_2\right)$; $S_E = S_E(z)$ is the lateral surface area of the electric heating element mounted in a gap of the peripheral annular channel; $k_l = 2 \Big/ \left[\dfrac{1}{\alpha_1 r_1} + \dfrac{1}{\lambda_m} \ln\left(\dfrac{\bar{r}_1}{r_1}\right) + \dfrac{1}{\alpha_2 \bar{r}_1}\right]$ is the average heat transfer coefficient, λ_m average coefficient of thermal conductivity of the material on the dividing wall, $\bar{r}_1 = r_1 + h_1$, h_1 is the wall thickness; α_1, α_2 is the average convective heat exchange coefficients on the inner and outer surfaces of this wall; $\rho_2, c_2, \bar{u}_2$, $\bar{t}_{02}$ is the average density, specific heat capacity, speed of the mixture flow rate, inlet temperature of the heated gas in the peripheral annular channel.

In the general case the density, heat capacity, average speed of the gas flow rate in the considered channels are close, $\bar{S}_1 < \bar{S}_2$. Taking this into consideration, assuming $q_E = 0$ we find

$$\bar{t}_1 \Box \left(1 - k_l \bar{S}_1\right)\bar{t}_{01} + k_l \bar{S}_1 \bar{t}_2,\ \bar{t}_2 \Box \left[\left(1 + k_l \bar{S}_1\right)\bar{t}_{02} + k_l \bar{S}_2 \bar{t}_{01}\right] \Big/ \left(1 + k_l \bar{S}_2\right).$$

It follows that if $\bar{t}_0 \Box\ \bar{t}_{02}$, then at the channel exit $(z = l_1)$ temperature of the powder-air mixture and thus temperature t_{pl1} of the powder particles will be

$$t_{pl1} = \bar{t}_{pl1} = \bar{t}_{01} + k_1 \bar{t}_{02}; \tag{4}$$

Temperature of the gaseous medium in the peripheral annular channel

$$\bar{t}_{2l1} = k_2 t_{02}, \tag{5}$$

where $k_1 = k_2 k_l \bar{S}_{1l}$, $k_2 = 1 \Big/ \left(1 + k_l \bar{S}_{2l}\right)$, $\bar{S}_{1l} = \bar{S}_1\left(z = l_1\right)$, $\bar{S}_{2l} = \bar{S}_2\left(z = l_2\right)$.

We can see that temperature t_{1l1} of the powder-air mixture at the outlet compared with the initial temperature increases by a value proportional to $\bar{t}_{02}$, heat transfer coefficient k_l, length of the first section l_1 and decreases proportional to the flow rate of this mixture. Respectively, gas temperature $\bar{t}_{2l1}$ compared to t_{02} decreases this is consistent with the existing views.

3. Second section of flow channel

At the second section heated powder-air mixture moves in the annular jet of the hot gas. By virtue of stability loss of the boundary line of the flows blurring of this boundary surface and mixing of the mediums occur. However, since the length of the section l_2 is relatively small, mixing is not taken into account here. It is assumed that the powder particles of the spherical shape with the same radius are under the uniform effect of radiation of a gas with the temperature approximately equal to $t_G \approx \bar{t}_{2l1}$, temperature of the inner surface of the tube is close to t_G. Under condition that powder particles in powder-air mixture can slow down or accelerate relative to the flow of a carrier medium, for example, under the action of an electrostatic field, coefficient of radiative-convective heat exchange of the particles with an environment can be written as

$$\alpha_G = \alpha_R\left(1 + \bar{\alpha}_C / \alpha_R\right), \tag{6}$$

where $\bar{\alpha}_C = \alpha_C\left(\bar{t}_{1l1} - t_p\right) \Big/ \left(\left(T_G/100\right)^4 - \left(T_p/100\right)^4\right)$; t_p, T_p is the temperature of the particle surface, α_C is the convective heat exchange coefficient of the particle surface, $\alpha_R = c_0 \varepsilon_c$ is the reduced emissivity, c_0=5.7 $\mathrm{W}\big/\left(\mathrm{m}^2 \cdot \mathrm{K}^4\right)$ is the blackbody emissivity, $\varepsilon_c = \varepsilon'_p \cdot \varepsilon'_G$ is the reduced emissivity factor of the system "body-gas"; $\varepsilon'_p = \varepsilon_p\left(1 + \left(1 - \varepsilon_p\right)\left(1 - \varepsilon_G\right)\right)$, ε_p is the emissivity factor of the surface material, $\varepsilon'_G = \left(\varepsilon_G - A_G\left(T_p/T_G\right)^4\right) \Big/ \left(1 - \left(T_p/T_G\right)^4\right)$ is the effective emissivity factor of the gas; ε_G, A_G is the emissivity factor, gaseous absorptivity ($A_G \approx \varepsilon_G$), for furnace gases $\varepsilon_G \approx 0.63$.

Having α_G, heat flow density acting on the powder particles per unit time can be calculated by empirical formula

$$q_p = \alpha_G\left(\left(T_G/100\right)^4 - \left(T_p/100\right)^4\right). \tag{7}$$

Further, assuming that powder particles are thermally thin bodies let us find their temperature in the second section after the time τ from the beginning of the flight

$$t_p = t_{pl1} + 3 q_p \tau \Big/ \left(\rho_p c_p R_p\right),$$

where ρ_p, c_p, R_p is the density, specific heat capacity and radius of powder particle, respectively.

In the case when particles velocity in the second section is close to $\bar{u}_1$, flight time in this section $\tau_l = l_2/\bar{u}_1$. Consequently, the particles temperature at the outlet of the second section

$$t_{pl2} = t_{pl1} + 3q_p l_2 / \left(\rho_p c_p R_p \bar{u}_1\right).$$

In the simplest case when $\alpha_C \approx 0$, $T_p \ll T_G = const$, $\varepsilon'_G \approx \varepsilon_G$, heat flow density $q_p \approx c_0 \varepsilon'_p \varepsilon_G (T_G/100)^4$, calculation ratio to estimate the temperature of particles material has the form

$$t_{pl2} \approx t_{pl1} + 3c_0 \varepsilon'_p \varepsilon_\Gamma l_2 (T_G/100)^4 / \left(\rho_p c_p R_p \bar{u}_1\right). \tag{8}$$

Using these relations we estimate the length l_1 of the first and l_2 of the second sections provided that powder material is polyethylene average thermal and physical characteristics are as follows: density $\rho_p = 10^3$ kg/m³, coefficient of thermal conductivity λ_p=0.35 W/(m·°C), specific heat capacity $c_p = 2.4 \cdot 10^3$ J/(kg·°C), melting temperature t_P=160 °C, decomposition temperature 290°C, average particles diameter $d_p = 10^{-4}$ m.

Assuming that the powder sprayer has a relatively small sizes and is intended for manual spraying the powder mass flow rate $G_p = 3 \cdot 10^{-3}$ kg/s, mass flow rate of air that contains the powder $G_{B1} = 5 \cdot 10^{-5}$ kg/s, mass flow rate of air in the outer annular channel $G_{B2} = 1.6 \cdot 10^{-4}$ kg/s. Cross sectional area of the central annular channel $S_1 = 1.9 \cdot 10^{-4}$ m², outer – $S_2 = 8 \cdot 10^{-4}$ m²; average speed of flow rate, respectively, $\bar{u}_1$=0.25 m/s, $\bar{u}_2$=0.15 m/s.

Inlet temperature of powder-air mixture $\bar{t}_{01}$=100°C, air inlet temperature in the outer channel $\bar{t}_{02}$=20 °C, heat flux into the air from the helical electrical heating element $q_E = 2 \cdot 10^3 / l_1$, W/m.

On condition that wall of the central tube is ceramic, its thickness h_1=0.003 m, λ_m= 0.5 W/(m·°C), radius r_1=0.008 m, taking into account the twist of powder-air mixture in it, as well as in outer annular channel [3-5], the intensification of heat exchange due to additional turbulization of the flows [6,7], the presence of impurity [8] we find: α_1=13.7 $\text{W}/\left(\text{m}^2 \cdot °\text{C}\right)$, α_2=3.25 $\text{W}/\left(\text{m}^2 \cdot °\text{C}\right)$, $1/\lambda_m \ln\left(\bar{r}_1/r_1\right) = 0.64$ m·°C/W, k_l =0.054 W/(m·°C).

As a result the calculated temperature of the heated air at the outlet of outer annular channel $\bar{t}_{2l1}$=440 °C, the length of the first section whereby inlet temperature of the powder-air mixture $\bar{t}_{1l1}$=160 °C, is l_1=0.16 m.

It should be noted that, this length can be reduced by increasing the temperature $\bar{t}_{01}$, increasing the power of a heating element, increasing the convective heat exchange coefficients α_1, α_2, increasing the coefficient of thermal conductivity of the wall material λ_m, decreasing the powder-air mixture flow rate.

Since it is necessary that at the outlet of the second section the powder particles have been melted, temperature of the material would have been approximately 200 °C from (8) we will find the length of the second section l_2 = 0.09 m.

4. Conclusion

The proposed mathematical models of heating and melting the polymeric powder particles as they move in the flow channel of the spraying gun quite completely reflect the specificity of these processes, contain all the basic design and operating parameters, allow solving the part of the problems of rational designing and operation of the plants for applying the polymer powder coatings.

References

[1] Yakovlev A D 1987 *Powder paints* (Leningrad: Khimiya) 216

[2] Galimov E R, Zverev E V, Tukbaev E E, Galimova N Ya, Kuryntsev S V and Mukhin AM 2012 *Polymeric powder coatings for special purposes* (Kazan: Publishing house Ofset servis) 164

[3] Mitrofanova O V 2010 *Hydrodynamics and heat exchange of swirling flows in channels of nuclear power plants* (Moscow: FIZMATLIT) 288

[4] Tarasevich S E, Fedyaev V M, Yakovlev A B, Morenko I V 2012 Experimental And Numerical Investigation of Heat Transfer In Annula Channels With Flow Twisting *Proceedings of the ASME 2012 Summer Heat Transfer Conference HT2012* Puerto Rico (USA) Order No.: I891DV. 6

[5] Kutateladze S S 1990 *Heat transfer and hydrodynamic resistance* (Moscow: Energoatomizdat) 367

[6] Isachenko V P, Osipova V A and Sukomel A S 1981 *Heat transfer: College textbook* (Moscow: Energoizdat) 416

[7] Morenko I V and Fedyaev V L 2010 Influence of turbulence of a viscous fluid flow on the hydrodynamic characteristics and heat exchange of streamlined bodies *Problems of energy* **7-8** 36-45

[8] Morenko I V and Fedyaev V L 2014 Laminar nonisothermal flow of a viscous fluid with solid particles past a rotating circular cylinder *Journal of Engineering Physics and Thermophysics* **87**(3) 566–572

IOP Conf. Series: Materials Science and Engineering **86** (2015) 012034 doi:10.1088/1757-899X/86/1/012034

Enhancing performance of measurement of parametric sensors parameters

V Zamorskiy

Kazan Federal University, Naberezhnye Chelny Institute 423810, Naberezhnye Chelny, Mira av., 68/19, Russia.

E-mail: altairvid@mail.ru, VVZamorskij@kpfu.ru

Abstract. The way of improving performance of measurement of parametric sensors parameters is considered. The result is achieved by calculating the elements of phase-frequency characteristics of the measuring circuit during the transition process while excluding the influence of interfering components.

1. Introduction
In practice, during measurement regardless the method is not always possible to avoid commutation, and hence there will be transition processes due to the inertia of the electric circuits elements. During the transition process, the measured voltages and currents depend on all the elements of the scheme, including internal complex resistance of measuring devices, energy sources and lead wires. Therefore, measurements, i.e. a count of measured values is carried out during steady-state process.

Time of transition processes damping is different and determined by the parameters of specific measurement system. Almost always, this time is much longer than the reference frame time of the required parameters. This substantially reduces the performance of measurement.

2. Analysis of the problem and formulation of the problem.
There are ways to measure of the separate determination of the parameters of the bipolar electrical circuits on the instantaneous values of transition processes [1], [2]. This approach significantly reduces the measurement time. But it involves the use of multiple transducers the influence of which on the result is not considered, as well as the influence of the power supply and connecting wires.

There are ways that allow to identify uniquely the elements of the sensor according to its phase-frequency characteristic [3], [4]. However, they can be applied only at the steady-state output signals of the studied object. In addition the method does not take into account the effect of interfering elements: internal complex resistance of measuring devices, energy sources, connecting cables and contacts. This leads to a very large error. Therefore it is only used for identification and not for measurement

In [5] approach to the measurement of passive two-terminal at different frequencies is well founded. Its essence consists in the formation of array responses of multi-element two-terminal to sinusoidal measuring impact. They are obtained at different frequencies in the time domain x(t,s), by analog and digital conversion. Then, using computational tools make the transition to the frequency domain X (jω,s) or X (p,s). It is known , that

$$X(p,s) = \left[\frac{c_n p^n + c_{n-1}p^{n-1} + \cdots c_1 p + c_0}{c_{2n}p^n + c_{2n-1}p^{n-1} + \cdots c_{n+1}p}\right]^{\pm 1}, \tag{1}$$

wherein n – the number of independent contours of two-terminal, s = (s1, s2,…, sn) – two-terminal parameters vector. Coefficients ci are continuously differentiable functions of the parameters si, т.е. ci = ui(s).

«+1» in the exponent corresponds to the resistance function Z(p,s) , a «–1» – functions of the conduction Y(p,s) of two-terminal.

Components si - desired parameters are determined by solving a simple system of nonlinear equations ci = ui(s), derived from (1) using known mathematical operations.

Properties of this method are, first, the dependence of the measurement result of the parasitic parameters of the measuring circuit. Second, the ability to provide acceptable metrological characteristics only in the steady state so as these transformations do not take into account all the factors affecting the process.

In the device [6] the method for the four-element two-terminal measurement at two frequencies, analogous to [5]. Measurement is performed as well in the steady state. However, its property is that, due to changes in the topology of the measuring circuit and the use of linear fractional transformation during processing, the measurement result is independent of the parasitic parameters of the measuring circuit, as well as the frequency volume of the energy source.

This feature can be explained by a simplified example.

Let the measuring circuit is as it is shown at 'figure 1'.

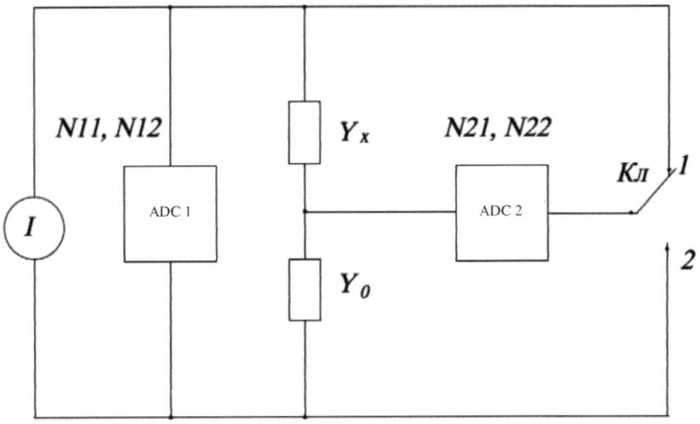

Figure 1. Method of compensation $Y_{ИСТ}$ and $Y_{АЦП}$.
I – current source; Yx – unknown conductivity; Y_0– exemplary conductivity; ADC 1 и ADC 2 – analog-to- digital voltage transducers; Кл – key, N11, N12, N21 and N22 - readings of the appropriate devices in the relevant tact.

Measuriments are carried out in two tacts. In the first tact the key Кл is in positon 1, and in the second tacta – in position 2. In the first tact ADC 1 counts N11, and ADC 2 - N21. In the second tact ADC 1 counts N12, and ADC 2 - N22.

Calculation of Yx is carried out by the formula:

$$Yx = Y_0 \frac{N_{11}}{N_{21}} \cdot \frac{N_{22}}{N_{12}} \tag{2}$$

Analyzing this circuit is easy to see that the count does not depend on $Y_{ИСТ}$ and both $Y_{АЦП}$ and their complex transmission coefficients.

3. Solution

To increase the performance is proposed to use a measurement method described in [5], during the transition process. And it becomes possible as for the impulse as for frequency measuring impacts by using the measuring circuit topology changes and apply extra fractional-linear processing described above.

A new method, the essence of which is that it is necessary:

1. Generate response arrays of multielement two-terminal to sinusoidal measuring impacts at different frequencies in the time domain x (t, s), through analog-to-digital conversion during the transition process (true for the single impulse impacts also).
2. Change the measuring circuit topology during the first phase so that extra fractional-linear processing will exclude the impact of complex finite values of internal resistance of the devices, power source and leading wires.
3. Then, using the computing means we make a transition in the frequency domain X (jω, s), or X (p, s).
4. Calculate the required components s_i,, solving simple systems of nonlinear equations $c_i = ui(s)$.

3.1. Example.

For example, consider a variant of strain-gage testing.

To increase the sensitivity and linearity the measurements are performed on an alternating current. However, with increasing frequency the effect of parasitic reactances of strain-gages increases. Strain-gage 'figure 2' is regarded as a complex resistance to intercontact capacitance *Cк* and inductance *Lд* ‘figure 2’. All measuring circuit should be considered, as it is shown in 'figure 3'.

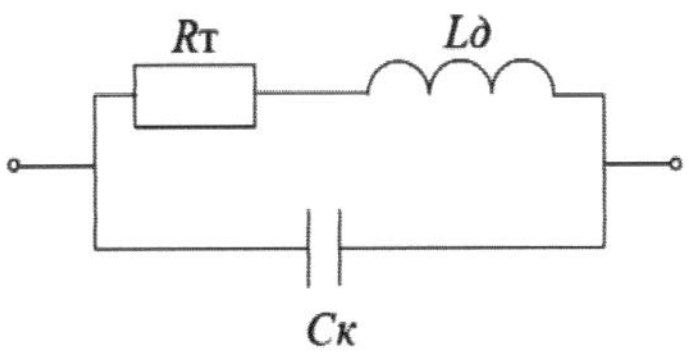

Figure 2. The equivalent circuit of the strain gauge

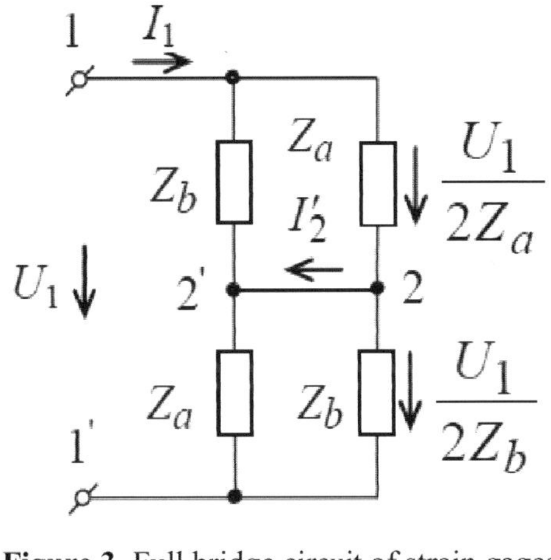

Figure 3. Full bridge circuit of strain gages turning on

Energy capacity of inertial energy accumulators that are placed on the rotating parts is small. Therefore, the power of measuring circuits is carried out pulsed, and secondary processing devices in the pauses are transferred into a sleep mode. This leads to the appearance of transition processes, which greatly reduces speed. The time spent for waiting for the transition processes attenuation are much longer than the time of signal processing.

To eliminate the effects of complex resistance measurement signal source and recording devices (АЦП) use a technique based on the proposed method.

Measurement is carried out as follows:

In the scheme 'figure 3' in the transition process, for example at an pulse action time functions arrays are received $U_1(t)$ и $I_2(t)$. With the help of the discrete Fourier transform we transfer to images $U_1(j\omega)$ и $I_2(j\omega)$.

From [7] it is known , that transfer function $H(j\omega) = \frac{U_1(j\omega)}{I_2(j\omega)}$.

In [8] is shown, that transfer function of bridge circuit 'figure 3' is calculated as

$$H(j\omega) = \frac{\sqrt{Z_b} - \sqrt{Z_a}}{\sqrt{Z_b} + \sqrt{Z_a}} \tag{3}$$

and $Z_a = Z_c \frac{1-H(j\omega)}{1+H(j\omega)}$, but $Z_b = \frac{Z_c^2}{Z_a}$, Zc – exemplary resistance equal to the characteristic impedance of the circuit образцовое $Z_c = \sqrt{Z_a Z_b}$, so, taking Za and Zc known and exemplary, we find

$$Z_a - Z_b = Z_c \frac{1 - H(j\omega)}{1 + H(j\omega)} - \frac{Z_c^2}{Z_a} \tag{4}$$

The resulting value (3) is used to determine the value of mechanical loads or displacements.

Application of the proposed methods will significantly reduce the measurement time by reducing the waiting for transition process damping.

4. Discussion of the results

In electrical engineering it is known that

$$\dot{U}_{ij} = n_i \dot{I} T_{ij} = \frac{n_i}{\Delta_j} \sum P_{ij} \Delta_{ij}, \tag{5}$$

wherein $\dot{U}_{ij}$ - output voltage i-th ADC in j-th measurement tact ; n_i - voltage transfer ratio i-th ADC in any measurement tacts ; $\dot{I}$ – source current; *Tij* - system function of the measuring circuit, which is determined by the ratio of the value measured i-th ADC in j-th measurement tact, to the magnitude of the current source ; *Pij* - the value of the transmission path through i-th ADC in j-th measurement tact; Δij – algebraic complement of the corresponding transmission path; Δj - determinant of the measuring circuit in j-th measurement tact [7].

In the first tact the division of readings ADC 2 and ADC1 is carried out. For voltages which they count value Δj is the same, so when divided the determinants are reduced. Attitude Δik for different devices are different so they are determined by the product of the relationship n2 and n1 complex transfer coefficient of appropriate ADC to algebraic complement of the transmission path through it (Y_0 is determined). In the second tact the same relationship is received with the only difference that it is determined by the ratio n1 and n2 and the value of Yx. In the calculation of (2) the relationships of complex transmission coefficients are reduced. That is why the dependence of the result of measurement Yx from parasitic parameters is excluded.

To eliminate the influence of leading wire resistance the circuit in Figure 1 must be complicated. Yx and Y0 must now be three or four-poles elements. The number of switching elements [9] increases also.

5. Conclusion

Complexity associated with the use of digital processing retards the establishment of high-speed measurement instruments and systems. Development of technical means, the widespread use of computers and digital signal processing allow to overcome the difficulties associated with the measurement during the transition process. This will allow you to create simple, cheap, fast and sufficiently accurate measuring instruments.

References

[1] Melent'ev, VS Methods and tools for the separate determination of the parameters of two-element bipolar electrical circuits / VS Melent'ev, EV Kostenko // Proceedings of the higher educational institutions. Volga region. Technical sciences. - 2009. - № 2 (10). - P. 102-112.

[2] Melent'ev, VS, methods for measuring the parameters of linear electric circuits by the instantaneous values of several transition processes / vs melent'ev, a.v.tsapaev // News of Higher Education. Volga region. Technical sciences. 2008. №2. S.80-90. Access: http://cyberleninka.ru/article/n/metody-izmereniya-parametrov-lineynyh-elektricheskih-tsepey-po-mgnovennym-znacheniyam-neskolkih-perehodnyh-protsessov#ixzz3TWNsn0tA

[3] Malanin, VP Analysis and synthesis of equivalent circuits of the "capacitive sensor - water-oil emulsion" / VP Malanin, AM Chivokin // Measurement. Monitoring. Management. Control. - 2012. - № 1. - P. 15-19.

[4] Abramov, SV Application of mathematical and physical modeling in the construction of the measuring circuit of eddy current sensors for non-contact displacement measurement / SV Abramov, VP Malanin // Measurement. Monitoring. Management. Control. - 2012. - № 1. - P. 40-44.

[5] Bobylev, DA Approach to the construction of multi-purpose transducer parameters-ports / DA Bobylev, LP Bohr // Measuring equipment. 2014. № 12. C. 47-51.

[6] AS 1244598 USSR. A device for measuring the parameters of parametric two-terminal / VV Overseas, GI Sharon // Byul.- 1986. - № 26.

[7] Dolbnya, VT Topological methods of analysis and synthesis of electrical circuits and systems / VT Dolbnya. // Kharkov Vishcha School, 1974. - 145 p.

[8] Ziborov SR Synthesis of linear radio chains: Textbook / SR Ziborov. - Sevastopol, Publ SevNTU, 2013. - 92 p.

[9] Sharon G. Methods for measuring the parameters of parametric two-terminal of multipolar electric circuit of the "triangle" type, Scientific Library KiberLeninka: http://cyberleninka.ru/article/n/sposoby-izmereniya-parametrov-passivnogo-kompleksnogo-dvuhpolyusnika-mnogopolyusnoy-elektricheskoy-tsepi-tipa-treugolnik

Determination of the oscillation frequencies of corrosion defects finite element methods in order to develop methods of acoustic monitoring of pipelines

Yu V Vankov, Sh G Ziganshin, E V Izmailova, V V Serov

Kazan State Power Engineering University, 51 Krasnoselskaya str., Kazan 420066, Russian Federation

E-mail: yvankov@mail.ru

Abstract. In this paper discusses issues related to increasing the reliability of heat supply systems. To determine the diagnostic criteria for defects in the numerical modelling of pipeline components and determined the frequency ranges in which it is necessary to analyze the acoustic signals.

1. Introduction

Currently about 60 percent of the pipeline heating systems in the Russian Federation have outlived their usefulness. There is no possibility of heat supply organizations to replace a number of such pipelines. To ensure reliable heat supply way out of this situation is the use of nondestructive testing and technical diagnostics in order to timely address the causes of possible failures.

A large number of failures due to corrosion of pipelines, therefore, the development of new methods for the detection of corrosion damage to metal is an urgent task. It is now widely used methods based on the analysis of natural acoustic oscillations in various continuous media [1], [2]. When choosing a method of control is necessary to take into account the fact that the heat pipes are in operation are mostly underground and difficult to access for inspection. It is known [3], [4] that for solving such problems, acoustic control methods have advantages over other methods.

2. The method of monitoring the technical condition of pipelines

Research is devoted to the development of the method of control piping analysis of acoustic signals propagating along the transported product from fluctuations pipeline sections with reduced due to corrosion of the wall thickness.

To date, no significant defined informative frequency ranges of acoustic spectra pipelines, indicating the presence of a defect of the pipeline wall, its size, for loaded circulating fluid piping.

The complexity of identifying pipeline corrosion defects during the experiment set the task to develop methods of numerical modeling of pipelines with different corrosion defects.

In order to find in the acoustic spectrum pipelines frequency domain characteristic of corrosion defects in the software environment of ANSYS were calculated natural frequencies of pipeline sections with models of corrosion damage in the form of a round plate with different sizes.

Simulated pipeline sections (Figure 1, Figure 2) 1500 mm long with corrosion defects dimensions 50x50 mm, 100x50 mm, 150x50 mm, 200x50 mm, 250x50 mm, and different values of wall thinning.

Corrosion defects in the form of reduced wall thickness was modeled disc rigidly clamped along the perimeter. Tube material is steel density $\rho = 7860$ kg / m^3, Poisson ratio $\nu = 0{,}3$ and a modulus of elasticity $E = 20 \cdot 10^{10}$ N / m^2.

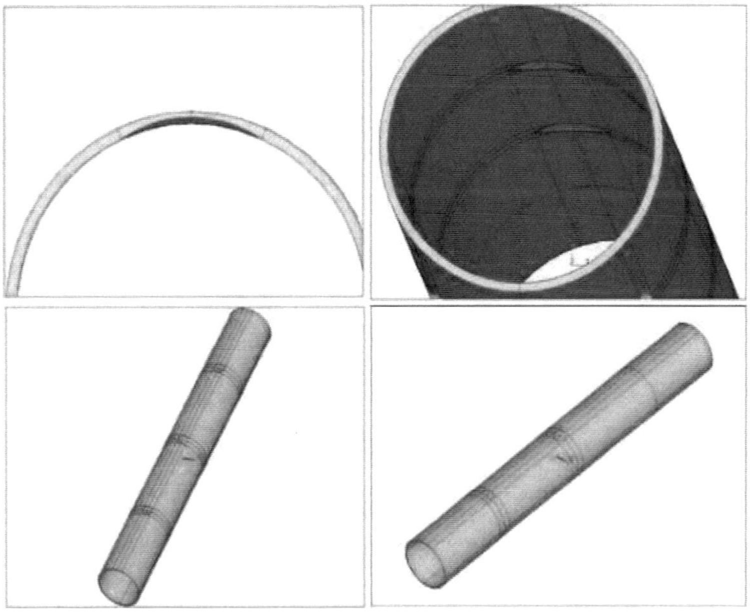

Figure 1. Finite element model of pipeline sections.

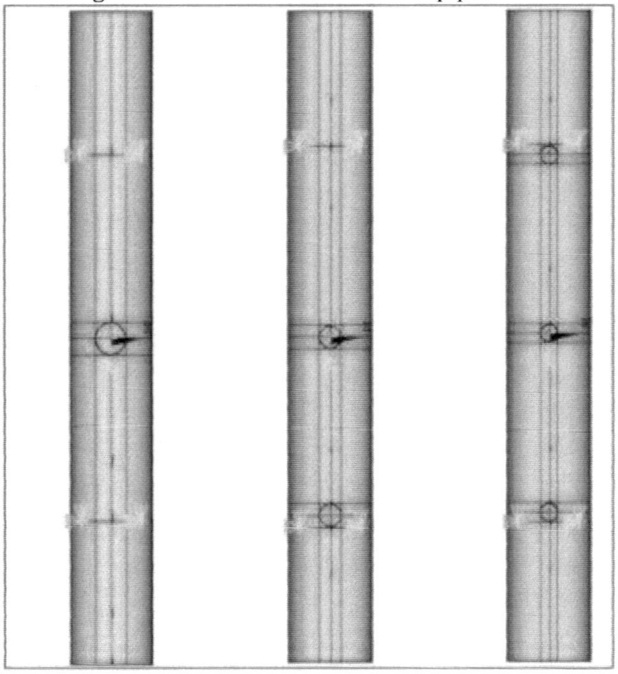

Figure 2. Finite element model of pipeline sections with the same area of corrosion damage but a different number and size of defects.

Figure 3 shows the dependence of the change the natural frequency of the defect model of the depth of penetration into the wall material. From the above graph it is clear that with increasing depth of corrosion damage occurs reduction in the frequency of different forms of oscillation disk.

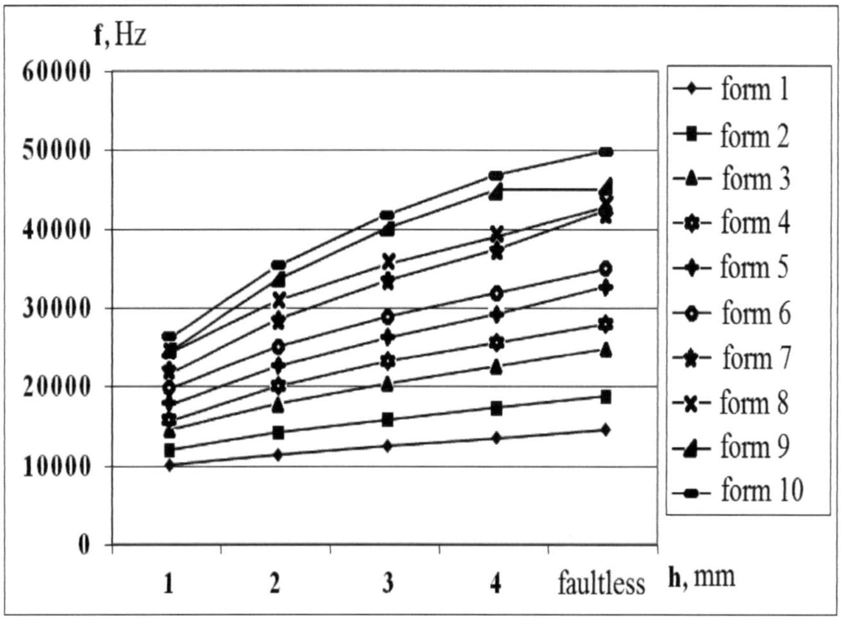

Figure 3. Changing the natural frequency of the model pipeline corrosion defects on its depth (disc diameter 50 mm).

3. Conclusions

As a result of numerical simulation to the following conclusions:

1. In the vibration spectrum faultless tubes are present primarily low frequency of 200 to 3000 Hz, and for tubes with defects increase characteristic oscillation frequencies up to 4500 Hz, and the presence of additional frequencies in the range from 6000 to 9000 Hz. The presence of small defects manifested in the higher frequency range.
2. It has been established that an increase in the number of defects in the same area affected frequency range of oscillation section of the pipeline increases. Extension of the range due to the fact that the defects, which are a kind of membrane, emit frequencies higher orders.

The work is performed according to the Russian Government Program of Competitive Growth of Kazan Federal University.

References

[1] Vankov Yu V 2003 Izvestiya Vysshikh Uchebnykh Zavedenij. Aviatsionnaya Tekhnika **(3)** 69-71
[2] Larionov V M and Nazarenko T I 2000 *Izvestiya Vysshikh Uchebnykh Zavedenij. Aviatsionnaya Tekhnika* **(4)** 69
[3] Belov E V, Vankov Yu V, Ivshin I V, Kochergin A V, Pervukhin D N, Tunakov A P 1996 *Acoustical Physics* **42(1)** 14-17
[4] Vankov Yu V, Kazakov R B, Pervukhin D N, 2004 *Zavodskaya Laboratoriya. Diagnostika Materialov* **70(3)** 34-38
[5] Vankov Yu V and Kondrat'ev A E, 2004 *Pribory i Sistemy Upravleniya* **(2)** 45-53
[6] Vankov Yu V, 2004 *Pribory i Sistemy Upravleniya* **(2)** 53-58

The effect of the size and position of transverse cracks in the natural frequencies of the rods

Yu V Vankov, Sh G Ziganshin, E V Izmailova, A R Zagretdinov

Kazan State Power Engineering University, 51 Krasnoselskaya str., Kazan 420066, Russian Federation

E-mail: yvankov@mail.ru

Abstract. In this paper discusses the issues related to the development of the acoustic method of free oscillations in relation to the diagnosis of cylindrical products. The results of numerical experiment to determine the oscillation frequencies of rods with defects such as "crack".

1. Introduction

Control methods based on the measurement of natural frequencies are sufficiently widespread to determine the physical and mechanical properties of different kinds of objects, materials and products [1], [2]. It is known that the acoustic method of free oscillations allows you to quickly, cost-effective resources to make decisions about the technical condition of controlled products [3], [4]. Review of publications showed that topical are issues of control of blades blanks of turbine engines, axles of railway cars, rods pumping units of oil wells, the pistons of internal combustion engines, etc. We solve the problem of the method of free oscillations with respect to the control of cylindrical products in the parameters of oscillations.

2. Theoretical analysis

The object of the calculations is cylindrical rods 300 mm long and 16 mm in diameter, mounted on two supporting knife located on the edges of the rod at a distance of 0.22 to its length. Determine the frequency and waveform defect-free rod and bars with cracks of various sizes and location.

Calculations were carried out using finite element method.

The problem of free oscillations provides for a harmonic law changes in the movement of time in the form of:

$$\{q(t)\} = e^{i\omega t}\{q\}. \tag{1}$$

The system of equations describing the free oscillations is of the form:

$$[M]\left\{\frac{d^2q(t)}{dt^2}\right\} + [K]\{q(t)\} = 0, \tag{2}$$

where [M] - stiffness matrix, [K] - mass matrix.

Substituting (1) into (2) and canceling the common factor, we obtain a system of linear algebraic equations:

$$[K]\{q\} - \omega^2 [M]\{q\} = 0 \tag{3}$$

IOP Conf. Series: Materials Science and Engineering **86** (2015) 012036 doi:10.1088/1757-899X/86/1/012036

From a mathematical point of view, equation (3) is a generalized eigenvalue problem whose solution is reduced to finding the spectrum of eigenvalues λi = ωi2 and their corresponding eigenvectors {qi}. In terms of the mechanics of the eigenvalues λi is the square of the frequency of free oscillations of ωi, and the eigenvectors {qi} characterize the shape of these oscillations.

Defects imitated incisions of different shapes with free edges, planes coincide with the planes of finite elements. Single defect depth was varied from 1 to 6 mm, the distance from the edge of the rod to the defect varied from 30 mm to 120 mm in increments of 30 mm (Figure 1).

Also studied two rods defects (Figure 2), the parameters of which are given in Table 1.

As an example of calculation results in Table 2 shows the frequencies of the first 28 modes of vibration of the rod without defect and defective four rods with different positions of transverse cracks.

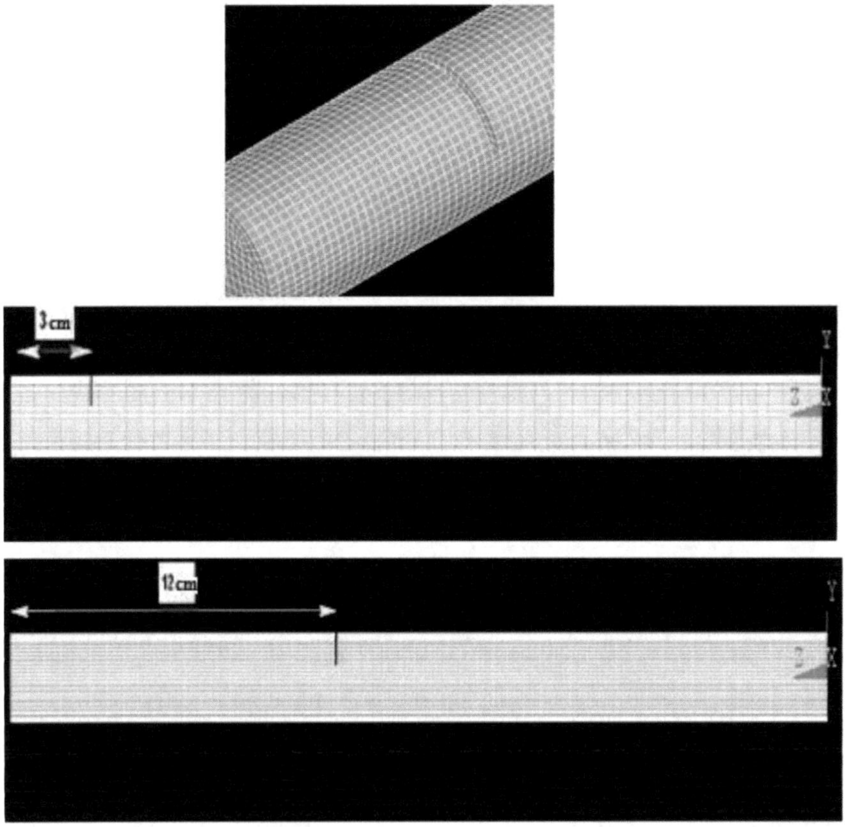

Figure 1. Model rod with a single defect rods.

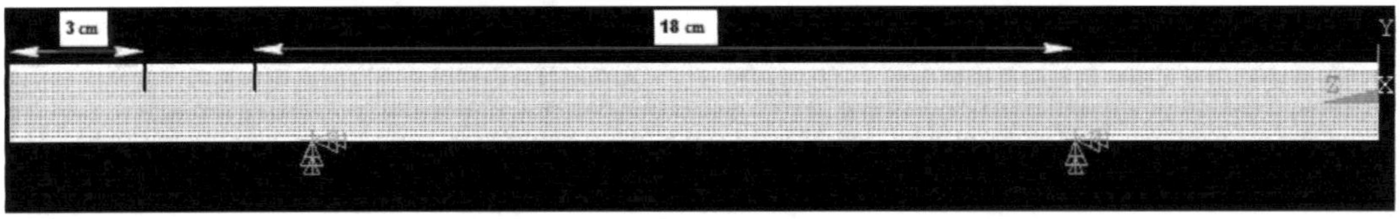

Figure 2. Model rod with two defects.

Table 1. Model parameters rods with two defects.

N	The depth of the defect	The distance to the defect from the edge of the first rod	Distance from the second defect to bearing supports
1	6 mm	3 cm	3 cm
2	6 mm	3 cm	6 cm
3	6 mm	3 cm	9 cm
4	6 mm	3 cm	12 cm
5	6 mm	3 cm	15 cm
6	6 mm	3 cm	18 cm

Table 2. The oscillation frequency of defect-free rod and four rods transverse cracks.

Number waveforms	The frequencies of the rod without defect	Frequencies of the rods with one defect			
		distance from the edge of the rod to the defect			
		3 cm	6 cm	9 cm	12 cm
1	794.4	794.87	792.8	789.32	785.73
2	944.9	1016.8	998.43	986.46	955.53
3	1302	1314.4	1300.4	1302.9	1309.6
4	1411	1396	1320.4	1340.1	1385.7
5	2201.1	2127.9	2030.8	2121.5	2127.3
6	2608.9	3028.6	2998.4	3025.7	3028.1
7	2997.2	3152.6	3152.6	3126.5	3107.4
8	3112.2	3270.1	3246.6	3271	3268.3
9	4468.6	4344.8	4331.2	4274.9	4185.1
10	6196.4	6210	6162.9	6167.7	6156.6

3. Conclusions

Results showed that the presence of a defect in the rod change over those frequencies antinodes waveforms which occur in the crack. Frequency, which coincides with the position of the crack node Current waveform, defect-free and have suitable rods substantially coincide.

Having a second defect leads to a shift in the oscillation frequencies corresponding to its size and position. The effect of each defect on the final product controlled oscillation spectrum is determined according to the principle of superposition.

To effectively detect the presence of defects in the product it is advisable to bring it at the same time the maximum number of frequencies and analyze all components of the spectrum.

The work is performed according to the Russian Government Program of Competitive Growth of Kazan Federal University.

References

[1] Vankov Yu V 2003 Izvestiya Vysshikh Uchebnykh Zavedenij. Aviatsionnaya Tekhnika **(3)** 69-71
[2] Larionov V M and Nazarenko T I 2000 *Izvestiya Vysshikh Uchebnykh Zavedenij. Aviatsionnaya Tekhnika* **(4)** 69
[3] Belov E V, Vankov Yu V, Ivshin I V, Kochergin A V, Pervukhin D N, Tunakov A P 1996

IOP Conf. Series: Materials Science and Engineering **86** (2015) 012036 doi:10.1088/1757-899X/86/1/012036

Acoustical Physics **42(1)** 14-17
[4] Vankov Yu V, Kazakov R B, Pervukhin D N, 2004 *Zavodskaya Laboratoriya. Diagnostika Materialov* **70(3)** 34-38
[5] Vankov Yu V and Kondrat'ev A E, 2004 *Pribory i Sistemy Upravleniya* **(2)** 45-53
[6] Vankov Yu V, 2004 *Pribory i Sistemy Upravleniya* **(2)** 53-58

IOP Conf. Series: Materials Science and Engineering **86** (2015) 012037 doi:10.1088/1757-899X/86/1/012037

Methods check the piping using a neural network

Yu V Vankov, Sh G Ziganshin, E V Izmailova, T O Politova

Kazan State Power Engineering University, 51 Krasnoselskaya str., Kazan 420066, Russian Federation

E-mail: yvankov@mail.ru

Abstract. The paper discusses issues related to the control of pipeline transport. To control the pipelines on the analysis of acoustic signals propagating along the transported product, it is proposed to use the neural network modeling. An example of implementation of the program in LabView environment for detecting defects of various sizes.

1. Introduction

Reliability and efficiency of urban heating and heat supply of industrial projects largely depends on the actual technical state of pipelines, and in particular for corrosion of pipes.

To ensure trouble-free operation of the pipelines necessary to have reliable information about the actual technical condition of the pipe, on the basis of which to make timely replacement of the "old" sites.

The above causes the relevance of diagnosing the state of non-destructive testing of pipelines. Solving the problem of diagnosing the state of corrosion of pipelines will solve the problem of reasonable prediction of their no-failure operation time and destination of their resources in accordance with the actual state.

It is well known [1], [2], [3], that the appearance of defects such as a crack, the change in geometric dimensions, physical-mechanical properties of the materials of construction leads to a change in the spectrum of natural oscillation frequencies of the product. Since during operation products defects may appear at different locations of the construction, their detection and identification by analyzing the spectrum of natural frequencies is difficult challenging. During the processing of acoustic signals faced with analysis of arrays a large number of frequencies obtained through experiments.

The use of computers makes it possible to handle this amount of data, but there are difficulties in creating a software algorithm of recognition and classification of defects. The aim of the study is to develop a method for monitoring the technical condition of pipelines in the parameters of acoustic signals based on neural network modeling.

2. Simulation of the diagnostic process

Neural network to diagnose pipelines has been implemented based on software complex LabView.

For the problem of defective pipeline network model is selected with training on error back-propagation algorithm. The network architecture is presented in (Figure 1).

The main program consists of several sub-programs, each of which performs a specific task. The first sub-program starts the task table reads files with recorded frequencies of pipes varying degrees of defects and a file with the frequencies of the tested pipe, and then the normalization of all vectors read. The second sub-program implements the method of learning by error back-propagation algorithm and classifies the defective pipe size of the defect. Sub-program is a neural network as a credential in it take the data from the tables of results of calculation of natural frequencies of membranes of different sizes, made by finite element method.

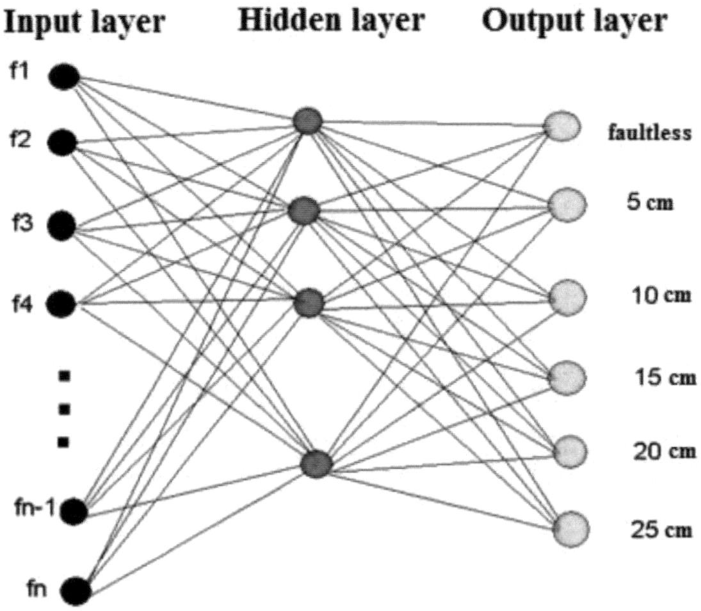

Figure 1. The architecture of the neural network training using the algorithm of back propagation of errors to diagnose the pipeline.

Established neural network consists of five output neurons, each of which characterizes the specific size defect. The program compares the training vector with diagnosable, and the output gives a result which of the classes (size) is defect. The input to the program of the samples submitted frequency sections of pipelines. The samples, in education, received frequency tube with known defects sizes 5x5, 10x5, 15x5, 20x5, 25x5 cm. As a result, the data output indicator lights on the front panel of the program (for the corresponding size of the defect, which corresponds to the maximum value of the likelihood and shows the numerical value of this probability). A typical report of the program is shown in Table 1.

Table 1. Probability distribution.

	The probability values of the test sample				
The true size of the defect	"Defect 5 cm"	"Defect 10 cm"	"Defect 15 cm"	"Defect 20 cm"	"Defect 25 cm"
5 cm	91,7	5,2	0,1	0,1	0,1
10 cm	5,2	90,1	4,2	2,3	0,9
15 cm	2,3	3,8	91,0	3,7	1,8
20 cm	0,5	0,7	4,1	90,5	7,2
25 cm	0,3	0,2	0,6	2,5	90,0

IOP Conf. Series: Materials Science and Engineering **86** (2015) 012037 doi:10.1088/1757-899X/86/1/012037

3. Conclusions

The results of the research can draw the following conclusions.

The method of classification of defects of pipelines using the technology of neural networks showed a high reliability of the results.

The results of the work helped to create a methodology to assess the technical condition of pipelines, including elements of solid mechanics, and a method based on the use of neural network analysis.

Analysis of experiments on the influence of defects on the oscillation parameters pipelines showed stable definition of the defect in the early stages of nucleation using neural network and confirmed the possibility of using the developed method for detecting defects of pipelines

The work is performed according to the Russian Government Program of Competitive Growth of Kazan Federal University.

References

[1] Vankov Yu V 2003 Izvestiya Vysshikh Uchebnykh Zavedenij. Aviatsionnaya Tekhnika **(3)** 69-71

[2] Belov E V, Vankov Yu V, Ivshin I V, Kochergin A V, Pervukhin D N, Tunakov A P 1996 *Acoustical Physics* **42(1)** 14-17

[3] Vankov Yu V, Kazakov R B, Pervukhin D N, 2004 *Zavodskaya Laboratoriya. Diagnostika Materialov* **70(3)** 34-38

[4] Vankov Yu V and Kondrat'ev A E, 2004 *Pribory i Sistemy Upravleniya* **(2)** 45-53

[5] Vankov Yu V, 2004 *Pribory i Sistemy Upravleniya* **(2)** 53-58

Structural and phase transformations in cast irons when implanted into the surface layers of nitrogen ions

E R Akst

Kazan Federal University, Naberezhnye Chelny Institute
423810, Naberezhnye Chelny, Mira av., 68/19, Russia.

E-mail: ev.akst@yandex.ru

Abstract. The results of the Mossbauer studies of graphite cast iron implanted with nitrogen ions are represented. We study the influence of the shape of graphite inclusions and modes of nitrogen ion implantation on the formation of in the surface layers of new phases and changes in the properties of cast iron.

1. Introduction

Ion implantation method, based on the implementation of the accelerated ions of different elements into the surface layers of solids, is quite effective, modern way to influence the properties of the materials. By means of ion implantation may be substantially changed the composition and structure of the near surface areas of irradiated materials that allows you to modify almost all properties of solids controlled by the state of the surface[1, 2]. Originally an ion implantation developed as technology to control the properties of semiconductor materials for microelectronics. Recently there has been a growing interest in studying the possibilities of ion implantation for the modification of surface-sensitive properties of metals and alloys. This was the basis for the formation of a new in ion-beam technology area - implant metallurgy. Compared with traditional methods of surface treatment of metal products (diffusion saturation, spraying, CTP), ion implantation has several advantages: a sharp reduction in the duration of the treatment process; increasing the solubility limits in the solid state and the independence of the formation of surface alloys from diffusion constants; no adhesion problems, as there is not a distinct boundary between the modified layer and the volume of the material; very little change in size of the part; opportunity to carry out selective processing of individual sections of it, and others. [3].

Progress in the field of implant metallurgy associated primarily with the development of fundamental ideas concerning the interaction of accelerated ions with the surface layers of complex metal systems. Of great importance are studies of structural phase transformations and mechanisms of the specific metallic materials properties change under ion irradiation. As for cast iron, it is the material relating to the category of highly complex, heterogeneous metal systems, as for the ion-beam modification prospects they are hardly been studied.

2. Object and methods of study

The object of study in this paper is graphite cast iron, whose surface is bombarded with nitrogen ions. The original structure of the samples was a ferrite-pearlite metal matrix with graphite inclusions of flake, vermicular and spherical shape. To assess the effect of the shape of graphite inclusions to the processes in the irradiated layers, samples for series of the experiments were chosen so that their chemical composition, the amount of graphite inclusions of graphite and density of distribution in the metal matrix were as close as possible. In this case the main variable factor is the total surface area of graphite inclusions per unit of the material volume, ie, density of graphite borders - minimal for cast iron with compact-spherical shape of graphite inclusions and higher for vermicular and flake graphite.

Irradiation of cast iron was carried out in accelerator IRA-3 by the stream of singly charged nitrogen ions with an energy of 40 keV in a dose range from 1 to 5 • 1017 • 1017 ions / cm^2, the ion current density of 50 mA / cm^2. The choice of irradiation modes is made based on the analysis of data from works [3, 10, 11, 12, 13 and others.] devoted to the study of the effect of nitrogen ion implantation on the mechanical and tribological properties of iron and iron-carbon alloys. Before and after cast iron irradiation the durability of samples was measured, and the microhardness of the surface layers also.

For registration of changes in structural and phase composition of the irradiated surface layers was used the nuclear gamma resonance (Mossbauer) spectroscopy (NGR spectroscopy). Mössbauer spectra were obtained in backscattering geometry by means of registration of internal conversion electrons and secondary characteristic X-rays. The choice of methods for measuring the spectra due to the fact that the samples studied cast irons are opaque to the Mossbauer gamma radiation and the traditional method of measuring of Mössbauer spectra of "with X-ray" for these objects is unacceptable. In case of registration of internal conversion electrons thickness of the probed layer (the escape electrons depth) is comparable to the average depth of the run of implanted in material nitrogen ions. Upon registration of the secondary X-ray a thickness of the probed layer by two orders of magnitude greater. This allows selective according to the depth researches - to receive information directly about the implanted layer and its adjacent layers with the registration of changes in the structure and phase composition [8, 9]. The possibility of a conventional X-ray analysis for solving such problems are severely limited, since the formation of the X-ray diffraction pattern the surface layer is involved, considerably exceeding in the thickness an area of intense radiation-induced phase formation.

3. Experimental results and discussion

Analyzing the results of Mössbauer studies, it may be noted that the Mössbauer spectra of all samples in the initial state are virtually identical and are representing a typical for the solid solutions of substitution based on α-iron superposition of sextets corresponding to atomic configurations with different numbers of impurity atoms of silicon in the immediate vicinity of Fe^{57}. The ratio of the intensities of lines in subspectra correlates with the chemical composition of the cast iron. In all spectra in addition to the dominant ferrite component it is also observed less intense lines sextet which is characteristic for cementite. In general, the nature of the Mossbauer spectra is in good agreement with the concept of the original structure, chemical and phase composition of the investigated cast iron [6, 7].

Irradiation of cast iron by nitrogen ions leads to a marked reduction in the relative intensity of the spectrum components, due to the presence in the ferrite silicon atoms. To the greatest extent this effect is seen in cast iron with flake and vermicular graphite at doses $D \geq 2 \cdot 10^{17}$ ions/cm^2. This phenomenon may be due to the following processes in irradiated irons. The bombardment of the surface layers of accelerated ions leads to the formation in the material the cascades of atomic displacements and accumulation along the trajectories of the implantable particles a large number of point radiation defects. In conditions of high radiation doses typical for such studies, the disordering processes (accumulation of defects) are compensated by their annealing processes (dynamic interstage and volumetric heat, due to local heating of the target). A significant part of the point radiation defects leaves the area of their generation and are sent to places of the effective capture (drain), ie to the grain boundaries, the free surfaces, etc. [4, 5]. As is known, the defects flow can entrain the impurity atoms having an atomic radius smaller than atoms of the solvent [3, 4]. This leads to a phenomenon of radiation-induced segregation, i.e. accumulation of impurity atoms in the drains. Obviously, in graphite cast iron the most efficient drain of radiation-induced defects are the boundaries of graphite inclusions, which may involve atoms of the dissolved in the ferrite silicon and implemented nitrogen atoms also. Since the density of drains (graphite borders) in cast iron with vermicular flaked graphite

IOP Conf. Series: Materials Science and Engineering **86** (2015) 012038 doi:10.1088/1757-899X/86/1/012038

and higher than in nodular cast iron, then these processes occur here more effectively. Another reason for the decline of the silicon content in the ferrite can be its selective spraying under the influence of the ion beam.

Starting with dose of cast iron irradiation $2 \cdot 10^{17}$ ions/cm^2 in their Mössbauer spectra obtained by the registration of internal conversion electrons, there are additional components that can be explained by the appearance of new phases in the alloys. In the spectra of the same samples, but obtained by the registration of the secondary characteristic X-rays, the additional components are not observed. This suggests that intensive phase formation caused by the implantation of nitrogen ions into the cast iron occurs primarily in the surface layers of a thickness from several tens to hundreds nanometers [6].

The parameters of the hyperfine structure (HFS) of additional components of the NGR spectra allow to connect them with nitrides ε-Fe_3N и γ'-Fe_4N. Cast iron with flaked graphite at a dose of $2 \cdot 10^{17}$ ion/cm^2 shares of subspectra ε-phase and γ-phase in the resulting spectrum is 6% and 12%, and at a dose of implantation $5 \cdot 10^{17}$ ion/cm^2 - 17% and 8% respectively. Shares of subspectra ε-phase and γ-phase in vermicular graphite iron at a dose of $2 \cdot 10^{17}$ ion/cm^2 are 16% and 7%, and at a dose of $5 \cdot 10^{17}$ ion/cm^2 - 18% and 12%. In nodular cast iron only at a dose implantation $5 \cdot 10^{17}$ ion/cm^2 are observed subspectra ε- and γ-phases and their proportions are 14% and 5%, and at a dose of $2 \cdot 10^{17}$ ion/cm^2 changes in the NGR spectrum are minor. Mathematical treatment of NGR spectra of cast iron with flaked and vermicular graphite also indicates the presence in the spectra of the paramagnetic and magnetic component with parameters HFS, distributed in a fairly wide range of values. The proportion of these components in alloys with flaked graphite with implantation at a dose of $5 \cdot 10^{17}$ ions/cm^2 is 18%, and in alloys with vermicular graphite - 13%. HFS parameters of these components allow you to associate them with the hexagonal ε-carbides and carbonitrides. Metastable ε-carbides in the ion bombardment conditions can be formed in atomic displacement cascades by the mechanism of atomic mixing [3, 4]. The most likely place for the formation of such carbides in the cast iron is obviously the border area of graphite and metal matrix. Regarding nitride ε-$Fe_{2+X}N$, then nitrogen and carbon are interchangeable in this compound and the concentration of these elements can vary here within wide limits ($0 < x < 3.2$). Changing the content of nitrogen and carbon in the carbonitride ε-$Fe_{2+X}(N,C)$ leads to a continuous change of HFS parameters of the corresponding subspectra HFS. For $x \leq 0{,}3$ magnetic ε-phase turns into a paramagnetic.

The results of measurements of microhardness and wear resistance of irradiated specimens correlate with the data of Mossbauer spectroscopy. The greatest effect of increasing the wear resistance and microhardness is observed in cast iron with flaked form of graphite inclusions, and the lowest - in nodular cast iron. In this case, the microhardness retained increased for all samples to a depth of not less than 50 microns, which is about two orders of magnitude greater than the thickness of the implanted layer [6, 7].

4. Conclusions

Analyzing the results and comparing them with data from the literature we can make the following assumptions:

1. Free carbon graphite of cast irons promotes caused by nitrogen ion implantation the formation process of secondary nitride and carbonitride phases as the relative proportion of these phases in the irradiated cast irons is higher than in pure iron and carbon steels with similar modes of implantation [10, 11].
2. Predominant phases formed in the cast iron by implanting nitrogen ions are nitrides, carbides and carbonitrides of ε-type. These phases have a hexagonal crystal lattice, similar to graphite lattice, and dissolved nitrogen and carbon in a wide range, which are used interchangeably herein. Apparently, under nonequilibrium conditions of ion implantation and in the presence of free carbon the formation of such phases is preferred in the kinetic and thermodynamic points of view. Leading role in the mechanism of phase formation processes may play a radiation-enhanced diffusion processes of impurity atoms to grain boundaries and ion mixing effect.

3. The formation of new phases takes place predominantly at the boundaries of graphite inclusions. These boundaries during irradiation cast irons by nitrogen ions play the role of effective drains (traps) of point radiation defects. Flow defects to drains entrains implanted into iron nitrogen atoms, which increases the likelihood of formation here of secondary nitrogen phases. Higher density of graphite boundaries in flake graphite cast iron causes a great number of secondary phases in these alloys. Density of graphite borders in nodular cast iron is the smallest, so the phase formation processes take place here less intensively.

4. The main reason for increasing the cast iron wear resistance and microhardness is the loss in the irradiated surface layers of nitride and carbonitride secondary phases. The greatest effect of improving the wear resistance and microhardness is observed in a flake graphite cast iron, and is accompanied by a higher content of said phases in the surface layers of the material. Another mechanism to improve the microhardness of cast iron may be an increase in the irradiated layers the density of point radiation defects that with the growth of its concentration in the are combined into three-dimensional clusters and dislocation loops, reinforcing the material. Increased microhardness of the samples at depths much greater than the value of average path of nitrogen ions in cast iron, can be explained by the phenomenon of radiation-enhanced diffusion of implanted atoms and radiation defects in the bulk of the sample.

References

[1] Gavrilova V A, Fazlyyyakhmatov M G and Kashapov N F 2013 *J. Phys.: Conf. Ser.* **479** 012010
[2] Galyautdinov R T, Kasparov N F and Luchkin G S 2002 *Inzhenerno-Fizicheskii Zhurnal* **75** 170–173
[3] F F Komarov Ion implantation into metals. - M .: Metallurgiya, 1990. - 216 p.
[4] Kirsanov V V, A L Suvorov, Trushin Y Radiative processes of the defect formation in metals. - M: Energoatomizdat, 1985. - 272 p.
[5] Akhiezer I A, Davydov L N Introduction to theoretical radiation physics of metals and alloys. - Kiev: Naukova Dumka, 1985. - 141 p.
[6] Akst E R, Romanov E S, Ivoilov N G Investigation of Phase Formation Processes In Nitrogen Implanted Iron-Carbonaceous Alloys // Surface. X-ray, synchrotron and neutron researches, 1995. - № 6. - P. 99-105.
[7] Vasilyeva E V, Usanova O. El-Bekai B.Sh., Lukyanenko E V, ER Axt Influence of the shape of graphite inclusions in the processes of phase formation and properties of cast iron subjected to ion implantation // Metally, 1996. - № 5. - p. 128 -134.
[8] Belozersky G N Mossbauer spectroscopy as a method for studying the surface. - M .: Energoatomizdat, 1990. - 352 p.
[9] Novakova A A, Kuzmin R N Mössbauer spectroscopy of conversion and its application. - M .: Publishing. Moscow State University, 1989. - 72 p.
[10] Longworth G, Hartley N Mossbauer Effect Study of Nitrogen-Implanted Iron Foils // Thin Solid Films. - 1978. - V.48. - № 1. - P. 95-104.
[11] Principi G, Matteazzi P, Ramous E, Longworth G Mossbauer Surface Study of Nitrogen-Implanted High Carbon Martensite // J.Mater.Sci. - 1980. - V.15. - P. 2665-2668.
[12] Kashapov L, Kashapov N and Kashapov R 2013 *Journal of Physics: Conference Series* **479** 012005
[13] Rauschenbach D, Hohmuth K Formation of Iron Carbonitride and Iron Carbide Phases During Nitrogen Ion Implantation // Crist.Res.Technol. - 1984. - V.19. - № 11 - P. 14255-1431.